A Hardware and Software Primer for Librarians

What Your Vendor Forgot to Tell You

Stephen Paling

The Scarecrow Press, Inc.
Lanham, Maryland, and London
1999

SCARECROW PRESS, INC.

Published in the United States of America
by Scarecrow Press, Inc.
4720 Boston Way
Lanham, Maryland 20706

4 Pleydell Gardens, Folkestone
Kent CT20 2DN, England

British Library Cataloguing in Publication Information Available

Library of Congress Cataloging-in-Publication Data

Paling, Stephen, 1965–
A hardware and software primer for librarians : what your vendor forgot to tell you / Stephen Paling.
p. cm.
Includes bibliographical references and index.
ISBN 0-8108-3572-X (cloth : alk. paper)
1. Librarians—Data processing. 2. Microcomputers. I. Title.
Z678.93.M53P34 1999
004.16—dc21 98-38882
CIP

∞™The paper used in this publication meets the minimum requirements of American National Standard for Information Sciences—Permanence of Paper for Printed Library Materials, ANSI Z39.48–1984.
Manufactured in the United States of America.

Contents

Introduction

Computer technology has reached into the workplace in ways that few people would have imagined. Computers have gone from duty primarily as word processors to being vital resources for the organization, storage, retrieval, and dissemination of information in a wide range of media. It is now a commonplace to compare the advances in computer technology to the invention of the printing press and movable type and to compare the Internet to a highway system on a global scale.

These technological advances have led to a great deal of uncertainty, perhaps outright anxiety, on the part of many librarians and administrators. People in the profession are being asked to use and analyze technology with which they may be quite unfamiliar. The rise of computer technology has produced a spate of books, but few of these serve as a manageably short, readable introduction to computer technology. This book is intended for librarians and administrators who feel a need to learn the basic vocabulary and concepts of computer use.

Very little experience on the part of the reader is assumed for this book. If you can perform routine tasks like word processing or checking e-mail, you should be comfortable enough to benefit from it. It is not intended to make anyone an expert, but rather it is intended to make library professionals more at ease in planning for, using, and discussing computer technology.

For the most part, specific software applications will not be discussed. Questions like "How do I lay out charts in my word processor?" are more specific than this book will cover. This is intended as an orientation and overview, a primer. General questions are often ignored by books which are written to teach the use of specific applications or meant for people who are already technically proficient and want to study a specific topic in depth. None of these gives a grounding in the basic terminology and ideas of computer use.

In general, specific prices and brand names will also be avoided, unless they add significantly to the discussion. Prominent brand names and prices change rapidly, and this book is not meant to be a consumer report that rates specific products. It is meant to make you a more informed consumer, which will help liberate you from dependence on salespeople and brand names in deciding which computer fits your needs.

The book begins with a breakdown of important hardware elements. It starts with central processing units (CPUs) and progresses through system buses, memory, and other basic components and peripherals. Hardware components constitute the foundation on which all computer use rests. Because understanding software often requires a discussion of the hardware and operating system software it requires, those explanations need to come first. After covering the basics of desktop computers, the book moves on to discuss local networks, the Internet, and the problems that go with networking. The final chapters discuss the advent of low-cost and network computers, two alternatives that have the potential to significantly change the way librarians and their patrons use computers, and finally, some helpful guidelines to use when it comes time to go shopping. Because these last two areas touch on so many other topics, they act as an effective summary and conclusion to the book.

Unavoidably, some sections of this book deal with the concerns of librarians more directly than others. For example, the chapter on CPUs consists mainly of technical background, while the chapter on low-cost and network PCs is framed very specifically for library use of these machines. This book can easily be compared with a book on printing and binding. Some aspects of bookmaking are specifically relevant to the needs of librarians and other people concerned with issues like preservation, while other aspects of the subject pertain to any type of book, no matter who is using it. Any book about computer technology will occupy a similar position. Some amount of generic background information must precede more specific discussion.

Before proceeding to the main body of the book, a few basic terms need to be discussed. One of the most basic areas of terminology involves the differences between the two main families, or platforms, of desktop computers: IBM-compatibles and Apples. The term *IBM-compatible* will refer to machines based on the various versions of MS-DOS and Windows, both by Microsoft, and using hardware based around the x86 line of CPUs developed by Intel. These are also sometimes referred to as IBM clones or simply PCs. The term *Apple* will be used to refer to computers made by Apple and other licensed manufacturers, built around a family of CPUs originally developed by Motorola, and running various versions of the Macintosh Operating System (Mac OS).

The details of what exactly constitutes a platform will be discussed later in the book, as well, but these basic differences should be kept in mind from the start. *Platform, CPU,* and *operating system* are all defined in the glossary for readers who want more exact definitions before beginning their reading.

It should also be noted that the term PC, often used to refer only to IBM-compatibles, will be used in its more general sense of personal computer to avoid confusion. Since both IBM-compatibles and Apples are personal computers, PC will refer to both. Some readers will no doubt note that the term IBM-compatible is no longer a strictly accurate term, since that platform has widened to include standards developed by many companies. But the term remains in normal use and will be used in this book.

The computer terminology and recommended specifications in this book serve three purposes. First, they provide an orientation to library and education professionals who feel unfamiliar with computers. Second, they can serve as basic benchmarks when deciding whether to stay with existing computers, upgrade them, or purchase new ones. Finally, they will help in the process of soliciting bids for technology projects. Many libraries and schools are forced to go with the lowest bidder when purchasing. Writing the correct specifications for computer equipment can help ensure that the lowest bidder will still provide acceptable technology. Budgetary, technical, personnel, and patron service issues all run together. Effective librarians and administrators must understand this part of their professional lives as well as any other.

If computers, and the jargon attached to them, seem less alien when you are finished, then this book will have succeeded.

I

COMPUTER HARDWARE AND COMPONENTS

1

CPUs and Motherboards

Definition

The *CPU,* or *central processing unit,* is the heart of a computer. It is a silicon *chip* that processes software instructions and turns them into action that you can see on your screen, at your printer, and so on. No other single part of a computer has as large an effect on performance, and the type and quality of CPU in a computer are among the most important factors to consider when measuring a machine's performance.

Chips, or *microchips,* are made of layers of extremely thin silicon wafers on which a series of electrical circuits is etched. The most important parts of these tiny circuits are the *transistors,* which act as tiny on/off switches. The pattern of on/off switches is the code the computer reads to execute commands, in much the same way a person might read the dots and dashes of Morse code. Chips can fill a variety of roles, including storage, but a CPU is a specialized type of chip called a *microprocessor* because it is specially configured to process commands sent to it by the user or by a piece of software.

Unfortunately, the term *CPU* is used in several overlapping ways that can confuse newcomers to computing. The computer itself—the case that contains the disk drives and other components—is often called the CPU. This use of the term *CPU* works in casual conversation, but is, strictly speaking, inaccurate. This book uses the term *computer* instead, and *CPU* will refer only to the computer's main processing chip.

The CPU resides on the *motherboard,* the main *circuit board* contained in the computer. A circuit board is a thin, flat piece of fiberglass to which various chips and components are attached. To give some idea of scale, a motherboard is roughly 1/8″ thick, and it is usually slightly longer and wider than a piece of notebook paper. Some of the attached components are removable, some are not. A number of these components will be discussed later, but for now we are only concerned with the CPU, the presence of which defines a motherboard.

The workings of the CPU and motherboard could be discussed in highly detailed, technical terms, but it will suffice to know that the CPU is the principal microprocessor in a computer and that the motherboard holds the CPU and other key components, like memory chips, which will be discussed later. The motherboard and CPU together make up a large percentage of a computer's price, and these two components help distinguish different models of computers.

Performance

The single most important measure of a CPU's performance is speed. This is expressed in *MHz* (*megahertz*), or millions of *cycles per second.* During each cycle, sometimes called a *clock cycle,* the CPU performs one or more operations. An operation is a very basic function, like shifting a piece of data or performing a simple arithmetic operation.

Newer processors can perform more than one operation, or instruction, per cycle, and thus some would claim that clock speed is a deceptive measure of a CPU's speed. These people cite *m.i.p.s.* (pronounced *mips*), or *millions of instructions per second,* as a more accurate measure of the processor's speed. Whichever is more accurate, chip producers and computer makers tend to emphasize the chip's clock speed, and information about a CPU's m.i.p.s. capability can be hard to find. The clock rate in MHz should be taken as the best measure of a chip's speed, and it can never be too fast. Software and operating systems are becoming increasingly complex, requiring faster and more powerful CPUs.

CPU Types

Two main families of CPUs are used in desktop computers: the *x86* line of chips, which power IBM-compatibles, and *PowerPC* chips, which are used principally in Apple computers. The current models in both families are powerful processors that dwarf the capabilities of their predecessors. For the average user, the technical difference between the two families is of little consequence, but because of their history, and the difference between IBM-compatibles and Apples, they need to be discussed separately.

x86 CPUs

IBM-compatible computers are far more numerous than Apples, and as a result the x86 line of CPUs is also much more common than its competitor. The name *x86* refers to the way in which the CPU model numbers were originally derived. The first chip in the series was the 8086, followed by the 80286, 80386, and so on. Users and marketers shortened these names to 286, 386, and eventually 486. Thus the term x86 came to be the generic term for this line of CPUs.

This system of naming broke down after the 486. Intel, the largest manufacturer of x86 chips, broke the mold with what would have been the 586 and called it the Pentium instead. Since then, Intel has released the Pentium Pro and the Pentium II, two powerful successors to the Pentium. The combination of Intel CPUs and the Windows operating system by Microsoft has become so ubiquitous that the combination is sometimes referred to as the *Wintel* platform.

Intel recently introduced an inexpensive chip, named Celeron, aimed specifically at computers running entry-level applications. Celeron's capabilities fall between those of the Pentium and Pentium II,[1] but it remains to be seen whether this chip will gain a significant share of the market. Its success may depend on the success of low-cost or network computers, which are discussed in a later section.

Intel is not without competition in the x86 chip market, however. Other companies, notably Cyrix and AMD (Advanced Micro Devices), have come out with high-performance, inexpensive competitors. These alternative CPUs deliver approximately the same level of performance as Intel's x86 chips and loosely follow the x86 naming scheme.

While the principal measure of CPU performance is speed, it is not the only measure. The 386 and 486 models were subdivided into the SX and DX series, based on the presence or absence of a *floating point unit,* or *f.p.u.* A floating point unit processes certain types of complex mathematical functions, taking part of the load from the main CPU and speeding overall processing. The DX designator indicated the presence of an f.p.u. *Math coprocessors,* or separate floating point units, could be added to some of the SX motherboards. Starting with the Pentium, all x86 chips have an f.p.u., but this old difference between models is still worth remembering.

All 386 and 486 chips should be considered obsolete. A 486DX running at 100 MHz or faster can offer approximately the same performance as some early Pentiums, but will not keep up with the rapid development of software. Word processing, *CD-ROM* searching for textual information, and so forth can be done on an older machine, but even those functions can strain an older CPU's capabilities. As library staff members and patrons increasingly use graphics-intensive Web pages and other demanding applications, these older chips will yield increasingly unsatisfactory performance, if they can run newer software at all.

Another boost in CPU performance, *MMX technology,* recently entered the market as an enhancement to Pentium and higher CPUs. First marketed by Intel, MMX increases the processing speed of *multimedia* programs. Multimedia refers to the use of more than one medium by a program; e.g., pictures and text, or sound with pictures and text. Intel's competitors are producing compatible technology for their CPUs. MMX technology adds a very modest amount, approximately $50–$100, to the cost of a computer. The added performance boost is worth the price, because increasing numbers of programs incorporate multimedia elements. A Pentium running at 200 MHz or higher, with MMX technology, should be considered the minimum requirement for a new system. MMX may soon become standard, removing that consideration from the decision.

Computer use can be divided into a number of categories for convenience. Entry-level applications perform basic tasks like word processing, spreadsheets, and e-mail as well as simple Internet and OPAC use. Moderate and high-end applications would include applications for statistics, graphics and animation (e.g., for use in online bibliographic instruction), and the viewing of complex Internet resources that make use of newer technologies. Power use refers to high-end design work, such as engineering, drafting, or movie-making, and most of those applications would be outside the needs of most libraries and classrooms. Table 1.1 summarizes the general classes of x86 CPUs and the level of use to which they can be put.

Intel's latest CPU, code-named Merced, is said to represent a significant

Table 1.1. CPU Models and Suitable Uses

Model	*Speed*	*Use*
386	Any speed	Obsolete
486	Any speed	Obsolete
Pentium	60–100 MHz	Entry level
Pentium	100–200 MHz	Entry level to moderate
Pentium	200+ MHz	Entry, moderate, high
Pentium Pro	150+ MHz	Entry, moderate, high, power
Pentium II	Any speed	Entry, moderate, high, power
Celeron	266 MHz	Entry level

break with previous x86 chip designs. Merced is due to be released in mid-2000, and details of its performance are not yet available. But whatever the details, it will feature a number of design innovations that may significantly boost the capability of the computers it runs.[2]

PowerPC CPUs

PowerPC chips are the closest competitors to the x86 family of CPUs, and they evolved in a slightly different way. Most of the original Apple computers were powered by the 68000 series of chips made by Motorola. These models progressed in speed, power, and model numbers in much the same way as the x86 family. The 68020 CPU succeeded the 68000, followed by the 68030 and 68040. All of those chips should now be considered obsolete and will be unable to run effectively much of the software being produced for Apples.

Rather than an evolutionary step, the PowerPC chip is meant to be a departure. Apple, IBM, and Motorola jointly created this family of CPUs, which were originally intended to run a variety of software both for IBM-compatibles and for Apples, but that never happened. IBM still sells its machines with x86 chips, and Apples are the only major platform using PowerPC CPUs.

Despite falling short of its original potential, the PowerPC family, and the Apple computers it powers, still enjoy a small but loyal customer base, especially in the education and desktop publishing markets. For this reason alone it is worthwhile to learn the taxonomy of the PowerPC CPUs. Unlike the x86 CPUs, the higher model numbers are not necessarily the best investment. CPU choice depends on the type of use the computer will get. Table 1.2 summarizes some of the capabilities and suitable uses for PowerPC CPUs. Apple has recently shifted its manufacturing focus and will probably place great emphasis on marketing machines featuring the G3 CPU. The G3 is specifically designed to work with the Macintosh operating system and software, and

Table 1.2. PowerPC CPUs and Suitable Uses

Model	*Speed*	*Use*
601	Lower speeds	Obsolete
601	Higher speeds	Entry level, obsolete soon
603	All speeds	Entry to high level
604	All speeds	High-end
620	All speeds	Power users, network servers
G3	All speeds	Entry to power level

Apple claims that this gives the G3 performance advantages not necessarily reflected in the raw speed of the processor measured in MHz.[3]

Summary

There are a number of circuit boards inside a computer, but the motherboard, which holds the CPU and other vital components, is the most important. The CPU is the brain, the main processing chip, of a computer. The motherboard and CPU largely define what separates different computer models.

The most commonly accepted measure of a CPU's performance is its speed, measured in MHz or millions of clock cycles per second. Other parameters, such as the number of transistors etched onto a chip or how many m.i.p.s. (millions of instructions per second) the CPU can perform, are also fairly good indicators of performance.

There are two principal families of CPUs used in desktop computers: the x86 family, originally made by Intel and used in IBM-compatibles, and PowerPC chips, used mainly in Apple computers. The use of the respective lines of microprocessors is one of the important differences between the two computer types. Performance varies widely within each family, and learning the basic taxonomy of each group is a good step toward being able to analyze the capabilities of a computer.

Glossary Terms

central processing unit
chip
circuit board
clock cycle
CPU
cycles per second
floating point unit
f.p.u.
math coprocessor
MHz (megahertz)
microprocessor
millions of instructions per second
m.i.p.s.
MMX technology
motherboard
multimedia
PowerPC
transistors
Wintel
x86

2

Expansion Cards, Slots, and Ports

Current computers derive much of their power and versatility from the fact that their capabilities can be changed and expanded in response to a user's needs. Some of these changes can be made by simple installation of new software, but other changes require the addition or replacement of hardware components. The relationship between these components can be complex, but a working knowledge of the components, their accompanying software, and the vocabulary that goes with them can go a long way toward making users more proficient with their machines. This kind of knowledge can also help in writing bids and understanding what makes some machines better than others. By the end of this chapter you will see that even without plans for expansion, knowledge of this part of computing is helpful.

Basic Terms

There are five basic terms to remember when discussing the expansion of a machine's capabilities: *bus, expansion slot, expansion card, port,* and *driver software* (*driver* for short).

Bus

A *bus,* in concept, is quite simple. It is a circuit that allows one part of a computer to communicate with another part. As an example, the CPU uses various buses to communicate with peripheral devices, such as modems and printers, which will be discussed in later chapters.

Expansion Slots

Expansion slots, often called simply *slots,* bear an important relationship with the buses in a computer. These slots are physical sockets on the machine's motherboard which allow circuit boards to be plugged into a bus, making these smaller boards part of the bus circuit and enabling them to communicate with the motherboard, the CPU, and with each other. On the outside of the computer, the slots appear as tall rectangular openings on the back of the computer, some of which will be occupied by cards with various connectors, and others will be blocked with cover plates.

Figure 2.1 shows a typical arrangement of slots on a computer. The actual number of slots will vary from machine to machine.

Expansion Cards

Expansion card is the most general term used to describe the circuit boards that plug into expansion slots. These boards are almost always dedicated to a very specific function, such as a modem card that enables the computer to communicate with other computers over phone lines.

Expansion cards plug into slots using *card edge connectors.* These connectors are electrically conductive strips on the edge of the card that mate with connectors on the inside of the slot. Figure 2.2 illustrates how a typical card would be installed. The view is from the side, with the rear of the computer being to the right. The plate on the back of the card, which would house any external connectors, replaces the cover plate which would normally cover the slot.

The actual number and arrangement of connectors will vary depending on the type of card. Some cards will have a notch cut out of the card edge

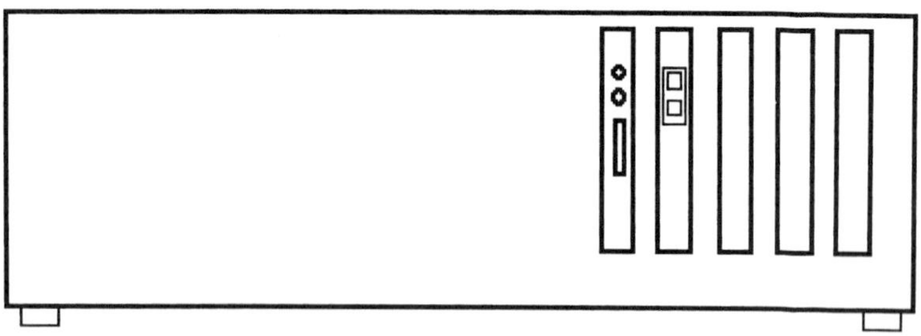

Figure 2.1. Expansion slots on the back of a computer.

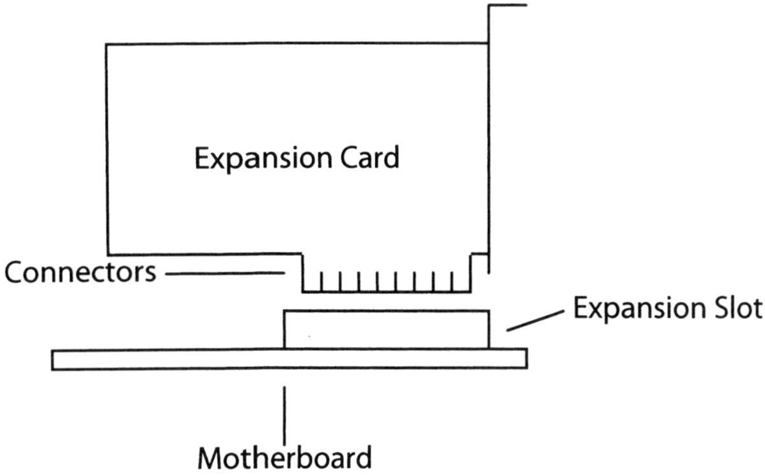

Figure 2.2. A typical expansion card and its slot.

connector that will correspond to a tab in the slot, ensuring that the card will not be inserted into the wrong type of slot.

Ports

A *port* is a physical connection to a computer,[1] such as the socket into which a keyboard or mouse is plugged. There are also internal ports, which work in conjunction with buses and cards, giving the card an assigned priority compared to other components discussed later.

Driver Software

Driver software, sometimes called just a *driver,* enables an expansion card or other hardware device to communicate with the rest of the system. Most cards will not work without the installation of this software. The software is usually included on a disk that comes with the piece of hardware.

These basic terms must be discussed in detail, though, before they begin to sound like more than interesting trivia. Even if a computer never receives an expansion card to augment its capabilities, it is important when buying a new machine to know what kind and number of slots are available. This tells the buyer a lot about the useful life the machine is likely to have, and how strong its performance is likely to be in comparison with other computers. Slots, ports, and expansion cards are evolving just like software, and a machine without a current hardware configuration at the time of purchase may well offer lesser performance than others on the market. In addition, there are many users who have decided to expand their computers' capabilities after all, only to discover that they do not have the right kind of slots for the expansion cards they need to use.

Buses and Slots

In practice, computers are rarely identified by the type of bus they have. Instead, only the overall number of expansion slots they have is indicated, and promotional material may be unclear as to which types of slots these are. But a good deal about a computer can be deduced by the type and number of slots it has, and it is worthwhile to learn the differences and to know which types are included in a new computer.

Buses and slots are more than different physical configurations for electrical connections. Different types of buses have different capabilities when it comes to the amount and speed of data moving across them, and sometimes different computers, which seem similar at first, will yield very different performance in real use. Computers have different model names and price points for many reasons, and the type and number of slots they possess is one of them.[2]

The 16-bit *ISA* (Industry Standard Architecture) slot is the most common in IBM-compatible computers, and it is the closest thing the industry has to a generic expansion slot. If a computer is said to have five expansion slots, for example, with no explanation of what type of slots, they are most likely to be 16-bit ISA slots. In this context, 16-bit refers to the width of the bus.

Bits and bytes, the basic units of computer storage, will be discussed in detail in the chapter on memory, but they merit a partial explanation here. Bits are small units of storage, and eight bits make up a byte, which is the equivalent of a single character or numeral in memory. So an 8-bit bus can move information eight bits, or one byte, at a time. This is faster than moving data a single bit at a time, though this slower method is still used at times. Moving from a smaller bus, like 8-bit, to a larger bus, like 16-bit, is referred to as increasing *bus width.*

The original ISA bus was an 8-bit bus. It is often referred to as an *XT bus,* since it was introduced in the IBM XT, an early personal computer. In 1984, IBM introduced the wider 16-bit ISA bus, the *AT bus,*[3] named after the IBM AT, a later personal computer. This wider bus, coupled with higher CPU speeds, created a significant increase in performance. Other bus types have been introduced since then, but most systems still have 16-bit ISA slots, though they may have other types as well. Almost all ISA slots in new systems are 16-bit.

There is a third, 32-bit version of the ISA bus, called *EISA,* or Extended Industry Standard Architecture. This type of slot is fairly rare, though, and there are not many cards available to fit it.

The *MCA bus,* short for Microchannel Architecture, was an effort by IBM to improve on the ISA standard. MCA came with some models of the IBM PS/2, but it never caught on. Owners of PS/2s, though, need to be sure that this is the type of card they are buying if they want to add an expansion card.[4] This card may need to be bought directly from IBM.

It was mentioned earlier that the CPU uses various buses. One of these is the *local bus,* which runs at a higher speed than some of the buses already mentioned. The most common of the cards to use this particular bus is the *VESA local bus,* developed by the Video Electronics Standards Association (VESA). It is more commonly referred to simply as a *video local bus,* or *VL bus,* because almost always it is used for the support of video cards, the cards that provide the signal to a monitor. It is common on IBM-compatible computers with a 486 chip, but is fading from prominence. For various technical reasons, the VL bus will probably go the way of the 486 as more advanced chips, like the Pentium or Power PC chips, continue to gain market share.[5]

The PCI bus is the current successor to the VL bus. PCI stands for Peripheral Component Interconnect. It is commonly used for video cards but can be used for other types of cards as well. The PCI bus has a 32-bit data width, and 64-bit versions are coming.

The most common configuration on new systems is a combination of 16-bit ISA slots and 32-bit PCI slots. The number of each will vary from machine to machine.

For users of laptops, the *PCMCIA slot* will be the most common. These slots allow the insertion of a credit card-sized expansion card, usually referred to as a *PCMCIA card,* or simply a *PC card.* Laptops and portables generally lack the room for normal expansion slots, and the PCMCIA slot is a substitute. These slots generally come in pairs, and no disassembly of the computer is needed to insert a card.

Another promising bus for delivery of video, *AGP,* or *Accelerated Graphics Port,* has been developed by Intel but has not been available long enough to be widely used. The AGP bus runs at roughly twice the speed of the PCI bus,[6] and shows great promise. However, the full selection of operating system support, hardware, and software is not in place yet, and AGP is not needed on most current computers. But it will probably gain popularity as wider support develops.[7]

Two other bus types, *Fire Wire,* or *IEEE-1394,* and *USB* (*Universal Serial Bus*), merit mention. Both buses are new and as yet not very common, but they may become so in the future. Fire Wire is, at least currently, used largely for digital video systems using camcorders and VCRs, and it features the ability to move the large amounts of data involved in video processing. The USB is more likely to become popular because it allows the use of large numbers of peripheral devices, up to 127, to be exact. But it is not clear yet if USB will gain wide acceptance.

The installation section later in this chapter explains how to identify common slot types.

Expansion Cards

One of the critical items to remember is that *a card must match its slot to work properly.* If a card is inserted into the wrong type of slot, a number of things may happen. If the user is lucky, the card simply will not work. If the user is unlucky, damage may result to the card, the computer, or both. This will be discussed in more detail in the section on installation, but it bears emphasis.

Expansion cards perform a variety of functions, and it is worthwhile to discuss the common functions for these cards, and the type of slot each kind of card is most likely to need. If someone anticipates adding an expansion card for a certain function, it is a good idea to buy a computer that will support the type of bus typically needed for that card.

Video cards, which along with monitors are discussed in a separate chapter, are cards that provide a video signal to the computer's monitor. Because of the increasing dominance of Pentium and PowerPC systems, most video cards currently on the market use a PCI slot. The local bus cards are disappearing along with the 486 chip, and any new system should have at least one PCI slot, preferably more.

Sound cards generally fit into a 16-bit ISA slot. It is important to note that sound cards are often referred to as 16- or 32-bit cards, but this does not refer to the expansion slot required by the card. It refers to the way in which the card formats sound. That will be explained in a later section devoted to sound cards.

Modems can be connected to a computer in a variety of ways. Internal modems, i.e., ones that go inside the machine as opposed to on the desktop, most often will plug into a 16-bit ISA slot. Internal modems designed for laptop computers will usually come in the form of a PCMCIA card and plug into a PCMCIA slot.

Internal ISDN modems will generally be 16-bit ISA cards, though PCI models are available. A technical note is worth mentioning here. Strictly speaking, ISDN modems are not modems at all, but rather network adapters. The distinction between the two will be discussed in the later section on networks.

Ethernet adapters, which can also be used to connect a computer to a local area network, can be found for a variety of sockets, with ISA and PCI being the most common.

Ports

Ports are similar to expansion cards in that they allow the connection of external devices, such as printers. Ports are generally built into the computer, though, and involve finding the correct cable to run between the device and the port, rather than the correct card for a particular bus. There are two common types of ports—serial and parallel—and each has distinct characteristics that make it suitable for different applications. The basic difference, though, is in the way they transmit data.

Serial ports transmit data one bit at a time, a very slow but reliable method of transmission. They are also known as *RS-232* ports, or *COM ports* (short for *communications port*). It is important to note that the original IBM PCs only supported two COM ports, and that IBM-compatibles with DOS 3.3 or later will support four. That will be important to remember later in this chapter. The other important feature of serial ports is that they are *bidirectional,* i.e., they can send as well as receive information from the computer.

Serial ports come in a variety of types. To give a few examples, the ports for mice and external modems are generally serial ports. This variety is too great to picture all of the types here, but the most common, a 9-pin connector, is pictured in Figure 2.3. A *pin* is the male, or protruding, part of a connector that mates with the female socket. Female sockets are still referred to by the number of pins they can handle.

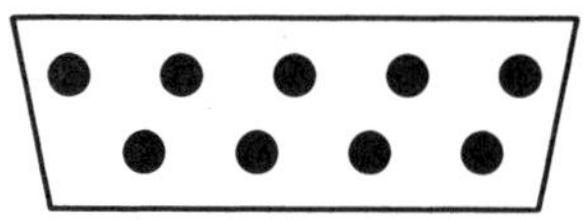

Figure 2.3. A typical 9-pin serial port configuration.

More important than the particular configuration of one connector or another, though, is the relationship serial ports bear to expansion slots and buses. Some cards, such as modems, take up a COM port *and* a slot. IBM-compatibles can have up to four COM ports, with two generally being external connectors and two being internal ports that work in conjunction with expansion slots.

Parallel ports, in contrast, transmit information eight bits, or one byte, at a time. This significantly speeds data transmission but also imposes certain limitations on devices connected to such a port, as we will see shortly. Most, but not all, parallel ports are *unidirectional,* i.e., they can receive information from the computer but cannot send any back.

The standard parallel port, or *Centronics port,* is a unidirectional port capable of sending data eight bits at a time. *Centronic* refers to the printer company that originally developed the standard connectors for these ports. IBM-compatibles support up to three parallel ports, referred to as LPT1, LPT2, and LPT3. LPT is short for Line PrinTer.[8] Most computers have just a single parallel port. This is usually adequate, but parallel ports can also be added as expansion cards if additional ones are needed.

Printers are the most common parallel port devices, but other parallel port devices are appearing. The most common of these is a scanner, which scans images and text into a form readable by the computer. Some of these scanners will require the installation of a second parallel port, others will not. Still others require the installation of a proprietary expansion card supplied by the scanner manufacturer. The best rule of thumb is to make sure you know all of a scanner's requirements before you buy it. This also holds true of other parallel port devices, such as external disk drives.

Parallel ports come with one very important caveat: they transmit information effectively only over short distances. After more than a few feet the signal can be interrupted or distorted, resulting in data loss. To be safe, a parallel port and cable should be used to communicate with a device only within a small area, such as a desktop or office.

Hardware Conflicts and Plug and Play

This chapter makes periodic references to conflicts between hardware devices. While these conflicts are becoming much less common, they are still an unpleasant reality in computing. A full explanation of what causes these conflicts and how to resolve them is beyond the scope of this book, but it is still necessary to discuss the basic factors that cause conflicts between devices.

One of the most important terms to remember in this context is *interrupt.* An interrupt is a signal sent from a hardware device to the CPU. It is a request for the CPU to interrupt normal processing and perform some action needed by the device sending the interrupt. Apple refers to interrupts as *traps.*

In principle, this may sound simple. When a device needs a piece of data, for example, it sends a signal to the CPU, which interrupts whatever it is doing to deliver the data, then resumes normal processing. But the variety of standards available for IBM-compatibles and their peripherals can cause problems with this process. These conflicts are also a problem in Apples but occur much less frequently than with IBM-compatibles.

In IBM-compatibles, most of the problems revolve around *IRQs*, short for *interrupt request.* IRQ is often used in two different senses. First, it refers to the actual requests sent to the CPU. It also refers to a numbered priority, zero to fifteen, assigned to each device that might need to send an interrupt. It is the equivalent of assigning each device a place in line, with the devices at the front of the line having higher priority. To make things more complicated, it is possible to accidentally assign more than one device to an interrupt, creating a conflict. These conflicts can manifest themselves in various symptoms. One or both of the conflicting devices may not work, or the first device to send a request may block the other device from any communication. Whenever new hardware has been installed, and problems develop immediately afterward, IRQ conflicts should be suspected.

Fortunately, this cursory view of hardware conflicts may be enough for many users, because the advent of *Plug and Play* is fast eliminating these configuration problems. The idea behind Plug and Play, or *PnP* for short, is that once a device is plugged into the computer, it is ready to be used immediately. Under a perfect PnP system, the operating system would communicate with each new device, and each device would configure itself automatically, choosing the proper port, IRQ, and any other resource needed. In practice, no PnP system is perfect, but they are improving. Windows 95 and the Mac OS both have PnP abilities, with the Mac OS being the better of the two at this particular task. Windows 3.1 does not have PnP capabilities.

There are some important caveats to go with Plug and Play, though. First, it is primarily designed to work with the PCI bus, though some ISA devices are also PnP compatible.[9] This is one more indication of the changes that are taking place in hardware design, and any system that does not have PnP-compatible slots is becoming obsolete.

If you are working under Windows 3.1, installation of hardware is more complicated than under Windows 95 or the Mac OS. Non-PnP boards have to be configured manually, usually by setting a series of *DIP switches* or *jumpers.* DIP switches (DIP is short for Dual Inline Package) are small switches usually mounted as a group on a circuit board. These switches set one or more parameters affecting the card, such as the speed at which it communicates with the rest of the machine, or which IRQ it should expect to be assigned. Jumpers perform a similar function, but they are more often found singly rather than in groups. Each jumper is a small, plastic-encased metal conductor used to connect two pins electrically. Again, this is usually used to set a particular parameter like IRQ selection. Figure 2.4 shows two sets of jumper pins, one set connected, and thus turned on, and the other set unconnected and turned off.

Different devices, such as sound cards or modems, customarily receive certain interrupts, and most boards come with their switches and jumpers set to this customary setting. There is no guarantee, though, that this default setting will work correctly for any particular machine. Some cards come with manuals that do a good job of walking a user through the process of resolving a conflict. But in most such circumstances, it is better to call the hardware manufacturer's technical support number. Because of the decreasing frequency and extreme complexity of these problems, it is not worthwhile for most users to learn how to fix them.

Before we leave this long explanation of boards, slots, ports, and their associated problems, there is another item worth mentioning. *SCSI* (pronounced *scuzzy*), or Small Computer System Interface, is one solution to connecting large numbers of peripheral devices to a computer. This has been left for last because SCSI combines elements of all the parts of this discussion. A SCSI controller is often an expansion card that plugs into an ex-

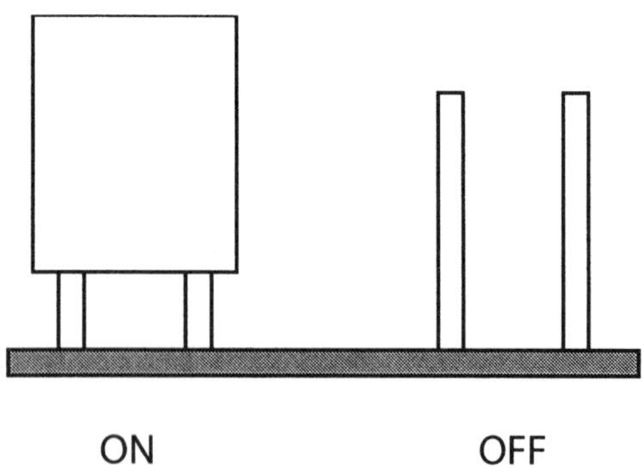

Figure 2.4. Two sets of jumpers.

pansion slot (though some can be built in), and it also serves some of the same functions as a port.

The most common reason to install a SCSI controller is the use of an optical scanner, since these are often built around a SCSI interface. One of SCSI's unique features is that it allows daisy chaining, which is the stringing together of multiple hardware devices on the same data path. In a typical daisy chain, you might see a SCSI scanner connected to the SCSI controller, and a SCSI printer connected to the scanner rather than to the computer itself. Data destined for the printer would pass through the scanner before reaching its destination. The final device in the chain must have a *terminator,* which tells the controller where the end of the chain is. There are various versions of SCSI, allowing the connection of anywhere from seven to thirty-two devices.[10]

The Universal Serial Bus, mentioned earlier, may replace SCSI. USB performs many of the same functions and allows for more devices and easier cabling. Future versions of Windows 95 and Windows NT are expected to support USB.[11]

All of this should make another thing clear: there are a large number of competing standards for card, slot, and port configurations, and we are far from being able to settle on a single unified standard. Plug and Play helps, but we are still a long way from eliminating the complications that come with the flexibility of modern computers.

Installation

All of the technical abstractions come together when a user actually installs an expansion card, whether that card is simply an interface between the computer and a device like a scanner or whether the card itself is the device to be used, such as a modem. But some users, understandably at this point, might ask themselves why, if a system comes with all the needed cards already installed, does the user need to learn about slots?

There are several answers to these questions. First, the term *expansion slot* is a misnomer. If a computer with five slots is purchased, but the machine already comes with a modem and video card, that means that two of the slots are already taken, reducing the total number of available slots to three. No matter how complete or wonderful a system seems when it is bought, never assume that its original configuration will remain adequate for the life of the machine. Computer capabilities are growing explosively, and new capabilities that often require new hardware make it important for a machine to have slots available for expansion. For example, a machine with no PCI slots is likely to become obsolete relatively quickly. A machine with a motherboard containing only ISA or VESA slots may be a bargain, but it will have limited expandability despite the apparent availability of slots.

Steps in Installation

Having said all of this, the mechanics of installing a card can be quite simple, at least when things go well. Following the proper steps can also help if you run into trouble and have to call technical support. The more you can explain about what you did, and in what order, the easier it will be for a support specialist to help you.

Finding a Proper Slot

Some manufacturers include a detailed diagram of the motherboard, which indicates exactly what type of slots are available and how many of each, but many machines do not come with this information. In that case, the first step is always to ascertain what slots are available for the installation of an expansion card. This requires opening the computer's *case,* or outer housing, and taking a look inside. Almost any computer will include documentation explaining how to open its case.

Once the computer is open, you will have to inspect the motherboard to determine which types of slots you have. The slots will generally run perpendicular to the back of the computer, terminating just behind the aluminum cover plates on the back of the machine.

Figure 2.5 shows part of an imaginary motherboard, with different slot types available. No real motherboard is likely to have all of these types. A real board would likely have a smaller number of slot types, but more than one of each. From left to right in the figure are an 8-bit ISA slot, a 16-bit ISA slot, EISA slot, VESA (VL bus) slot, and a PCI slot. The less common slot types are not pictured.

Finding a Compatible Card

Once you ascertain what type of slot is available, you need to buy an appropriate expansion board. This is often the simplest part of the process. Most catalog services that sell expansion cards list the type of slot each board requires, and most computer dealers can tell you the same information. In fact,

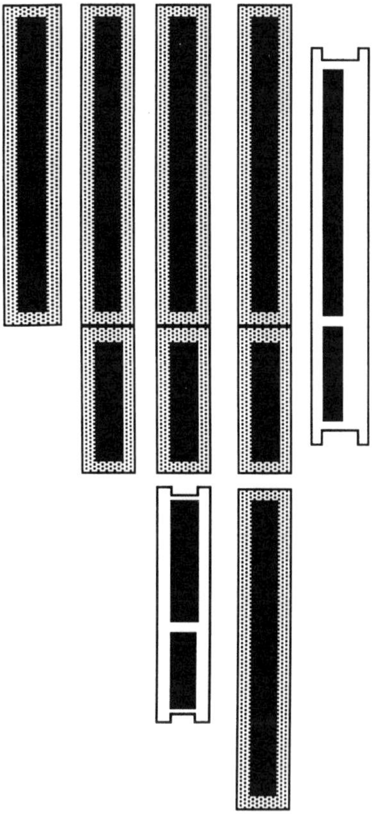

Figure 2.5. Common expansion slot types.

one of the ways you can identify good hardware vendors is by whether they can find this information. A responsible seller will understand its importance.

Physical Installation

The installation of the board is generally done before the installation of the driver software. Before beginning this step, always unplug the computer from its wall socket or power supply to avoid electric shock.

The first step in installation is to remove the metal cover from the expansion slot. It is usually held on by a single screw and comes out very easily. The screw will usually be reused to hold the card in place after installation, so keep it. The slot's cover should also be kept in case the card is ever removed. Incidentally, it is often convenient to install a card into a slot with empty slots on either side. This will give you more room to work. In most cases, there is no difference between slots of the same type.

Some cards will give instructions on setting jumpers or DIP switches, but most will have the cards set to an appropriate default setting. With Plug and Play cards, this should not be an issue.

Once the slot's cover is removed, insert the card into the slot. Before doing this, be sure to read any instructions that came with the card explaining how to shield it from static electricity. In some cases, inserting the card will take a surprisingly large amount of downward pressure. Some cards go in more easily if they are rocked gently from front to back. Do not rock the board from side to side. Once it slides firmly into place, secure it with the screw from the cover and reassemble the computer's case.

Software Installation

Software installation is often the step where any conflicts or problems will begin to manifest themselves. The installation directions that come with the card should always be considered the most important, and the instructions that follow here are only general guidelines.

Once the board is installed, restart the computer. What happens at this point depends on what operating system the computer runs and on whether the machine has any problems recognizing the board. In the context, *recognizing* refers to the computer's ability to see that a new card has been installed into one of the slots.

Under Windows 95 and the Mac OS, hardware recognition should be simple. Windows 95 should detect the new piece of hardware and walk you through the remaining steps in the installation. At some point it will probably ask for the disk holding the driver software for the component. This simply requires you to insert the disk into the floppy drive and tell the computer to proceed.

Under the Mac OS, software installation is even simpler. Generally, it is simply a matter of inserting the disk with the driver and then double-clicking on that disk's icon when it appears on the desktop. Sometimes the disk will contain an installation program that the user simply starts and runs, and other times the disk will contain files or folders that can be copied to the appropriate location on the machine's hard disk drive.

Windows 3.1 lacks some of the advanced features of Windows 95 and the Mac OS, making the installation of drivers potentially more difficult. A

Windows 3.1 driver will usually come in the form of an install program on a floppy disk. When the user starts the program, it will begin to install the drivers but will often ask the user a number of questions as it does so. These questions pertain to parameters like IRQs, and it is generally best to stay with whatever the default setting the program provides. Again, there is no guarantee that these default settings will work, but they are most likely the best settings. Once again, if all goes well, the user can restart the computer and the new hardware will work as intended.

If there is a problem at this stage, it can introduce the user to the most frustrating part of computing. A number of things can go wrong, and the best advice is to consult the manual, consult a friend who might be able to help, or to call tech support. But a few general rules can be used to ease the situation.

The most severe symptom would be if the computer failed to turn on, or failed to restart properly after installation of the card. In this case, remove the card, reinsert any card it replaced, and restart the computer. This will usually solve the problem. This sort of situation almost always requires outside help or the purchase of a new card. It can often be a sign that the incorrect type of card was bought.

Other problems of varying severity can occur. These usually involve faulty performance, or nonperformance, of certain features. The first resort should always be the manual. Under any circumstances, *keep track of what you've done.* Many of these problems will require a telephone call to tech support, and the more you can tell them about what you've done and what has happened (or not happened), the better off you'll be. It is unfortunate that this book can only provide such general advice, but the problems that can result are simply too numerous to cover in a single book.

Summary

Expansion cards are circuit boards made to fit into slots, connecting the card to a bus, in order to expand the computer's capabilities. Most expansion cards require driver software to communicate with the system. Ports are physical connections that allow the connection of external devices such as printers or external modems. There are also internal ports that are used by the expansion cards. ISA and PCI slots are the most common.

Even if a user never plans to add an expansion card, knowledge of cards, slots, and buses is important to understanding computers. The type and number of slots in a computer tell a great deal about the likely performance and future obsolescence of the machine. Buses affect many of a computer's input and output functions, which in turn have a large effect on the machine's overall performance.

Hardware conflicts can result as expansion cards compete for system resources. IRQs are the most common source of these problems, because more than one device can be assigned to each IRQ. These problems tend to be among the most difficult in computing, but they are becoming much less common as Plug and Play systems are developed.

Installation of expansion cards should be simple if there are no problems. Once the user determines what type of slot is available and obtains a compatible card, all that should be required is the insertion of the card and the installation of driver software. Often, when problems arise, the solution will involve changing the setting of a jumper or DIP switch, and many of these problems require a call to tech support.

Glossary Terms

accelerated graphic port
AGP
AT bus
bidirectional
bus
bus width
card edge connector
case
Centronics port
chassis
COM port
communications port
DIP switch
driver
driver software
EISA
expansion bus
expansion card
expansion slot
Fire Wire
IEEE-1394
interrupt
interrupt request
IRQ
ISA
jumper
local bus
MCA bus
parallel port
PC card
PCMCIA card
PCMCIA slot
Pin
Plug and Play
PnP
port
RS-232
SCSI
serial port
slot
terminator
trap
unidirectional
Universal Serial Bus
USB
VESA bus
Video Local Bus
VL bus
XT bus

3

Memory

After CPU type and speed, a computer's memory has the greatest impact on performance. The amount of memory and, to a lesser extent, the type of memory play a large role in determining how fast and how reliable a system will be.

The basic unit of storage in a computer is a *byte,* which represents one character, such as a letter or punctuation mark. Thus a *kilobyte* (kb) is approximately one thousand bytes, or characters, and a *megabyte* (MB) is approximately one million.[1] Each byte in turn is made up of eight *bits.* A bit is either a one or a zero, and the pattern of ones and zeroes in each group of eight determines which character is represented. Transistors were mentioned in the discussion on CPUs, and they play an important part in memory and storage as well. Each transistor on a memory chip represents a bit, and each transistor can be turned on or off like a switch. If the transistor is turned on, the bit is read by the computer as a one. If the transistor is turned off, the bit is read as a zero.

All the information in a computer, even nontext files such as graphics, is stored in this *binary* format. In the case of a graphics file, the program using that file simply interprets the pattern of bits and bytes differently than if it were looking for text. This is true of disk and other fixed storage media, as well.

RAM

The most important type of memory from the user's point of view is *RAM,* or *random access memory.* This is often referred to as *main memory* and should be thought of as the computer's work area. Whenever a user starts a program or opens a file of any type, that information is found in its normal storage place on a disk or other physical storage medium, and it is copied into memory for use.

One of the most important things to remember about RAM is that it is *volatile.* When the user turns the computer off, either intentionally or by accident, all the information in RAM is lost permanently. *Volatility* refers to this impermanence of storage in RAM. If the information has not been stored on disk, it is simply gone when power is cut to the computer. It is almost a rite of passage for new users to lose important information this way. Once a user does store a file on disk, the file is relatively secure even if the computer is turned off.

RAM comes in a variety of types. EDO RAM, SRAM, and DRAM (pronounced *ee-dee-oh ram, ess ram,* and *dee ram,* respectively) are a few of the types available for personal computers. There are differences in the way the circuits in each type are engineered, and some types of RAM can deliver information to the CPU faster than others. Although the typical user is unlikely to notice these differences in performance, they are worth mentioning only because they can confuse users who are not used to discussing computer hardware.

In most cases, the amount of RAM a machine has for its main memory is more important than the type of RAM. Newer types of memory, such as 100 MHz SDRAM, may change this rule of thumb, but for now they offer little performance gain on most machines.[2] Computer programs continue to increase in size, and these programs need larger amounts of RAM to run effectively. RAM for main memory is measured in megabytes, and 16 MB should be considered the minimum for any new computer, with 32 MB being preferable. Many older computers will also perform better if they are upgraded to this amount of memory. Adding RAM gives the computer a larger workspace, and this increase in capacity can have a dramatic effect on system speed and performance.

Sixteen megabytes of RAM is adequate for many library staff functions, such as word processing and e-mail, especially if the programs in use are strictly text-based and incorporate few graphic elements. That amount is not adequate for all tasks, though. Any machine used for graphics-intensive applications should have 32 MB of RAM. This speeds multimedia applications and graphics and also makes *multitasking* more practical. Multitasking increases the potential power of a computer by letting the user run more than one program at once, switching back and forth between the two as needed. For example, a user could search a database on CD-ROM and switch easily to an open Internet connection to search online as well. Switching between two programs becomes faster and more reliable with larger amounts of RAM. Multitasking is slower with smaller amounts of RAM and can even cause the computer to malfunction and lose data. Library use of graphics- and data-intensive programs will continue to increase, and 32 MB should now be considered the standard for new machines.

Extremely powerful applications, such as programs for full motion animation, often require 64 MB or more of RAM. Most library and classroom applications currently do not need this much RAM, but this will almost certainly change as program needs continue to grow. Any library staff member actively involved in the creation of Web pages, or any other technology-dependent activity, should probably have a machine with at least 64 MB.

Cache

Cache memory, often called *cache RAM* or just *cache,* is distinct from main memory and also plays a role in increasing system performance. Cache is a small amount of very fast memory in which the computer will store vital information for the program being run in main memory. This cached information is made up of parts of the program or data that must be accessed more frequently than others. Cache delivers this information to the CPU more quickly than normal RAM, speeding system performance. Cache comes in smaller amounts than main memory, usually no more than a few kilobytes, although some computers have larger amounts. In main memory, the type of RAM is not as critical as the amount, but with cache the opposite is true. It consists of specialized, more expensive RAM chips.

Cache can be added to many computers as separate memory chips, but it is now built into many CPUs. The amount of cache built into the CPU cannot be increased.

Upgrading Memory

Upgrading a computer by adding RAM is one of the easiest upgrades to perform and can have one of the largest payoffs in performance. Any user who follows a few simple guidelines will have little trouble adding memory to his or her machine.

Memory chips are not generic, and the most important consideration when adding memory is to purchase the right kind of RAM. A computer will usually take only the type of memory it came with, so it is vital to match this type. A good vendor should be able to determine what type of memory a computer needs, and many computer manufacturers have Web sites where that information can be obtained. Two examples of good vendor sites are www.ware.house.com/memory (MicroWarehouse) and www.kingston.com (Kingston Technology). Both sites offer information on selecting memory for machines from virtually any major manufacturer. It would be better, though, to note the type of memory a computer contains at the time of purchase. If the wrong type of memory is installed in a computer no harm should result. Most computers will simply ignore incompatible RAM.

Several important caveats should be remembered about RAM. The first caveat pertains to the form in which RAM is manufactured. Rather than being sold as individual chips, a number of RAM chips, usually two to six, are soldered onto a small circuit board a little shorter than an average person's finger. This circuit board, or *memory module,* plugs into the RAM slot.

There is more than one type of memory module. The more common of the two is a *SIMM,* or *single inline memory module.* SIMMs can be used singly. If a computer is equipped, for example, with two 8 MB SIMMs, for a total of 16 MB, another single 8 MB SIMM could be added to bring the computer's RAM to a total of 24 MB.

Many laptop, and some desktop, computers use *DIMMs, or dual inline memory modules.* DIMMs must be used in pairs, and this has implications for memory upgrades. DIMMs cannot be used singly, and many laptops have only two memory slots. So if a laptop with two slots has two 8 MB DIMMs, for a total of 16 MB, a single memory module cannot be added. If a user wants to increase the computer's memory to 32 MB, the two 8 MB DIMMs must be discarded and two 16 MB DIMMs added to raise the memory to 32 MB. The two original DIMMs are of no use to the user, and the user must also pay the extra money for the full 32 MB of RAM. DIMMs offer enhanced performance when compared to SIMMs, but at a higher cost. Computers that use DIMMs also tend to have a lower total capacity for RAM. These computers should not be ruled out by any means, but anyone seeking computer equipment on a limited budget should consider the increased cost.

The other caveat is that a computer will only have a certain number of *slots,* or *sockets,* into which RAM can be plugged. These are separate, smaller slots than expansion slots, and the two should not be confused.

RAM slots are almost always on the motherboard. Each slot will hold up to a certain amount of RAM, and this amount will vary widely, usually ranging up to a few dozen megabytes. If a motherboard has four RAM slots, each capable of holding a 96 MB SIMM, the machine's maximum RAM capacity is 384 MB. That would be a generous amount for the current needs of almost all users, but never assume that a computer will hold that much RAM. Almost any computer, even a laptop, now will hold over 100 MB, but this specification should always be checked before purchase. Desktops can routinely hold 300–400 MB.

ROM

ROM, or *read-only memory,* differs from RAM in several ways. The principal difference is that ROM can only be read. It cannot be written to under most conditions, making it nonvolatile. This gives rise to the second difference. ROM usually holds specific types of system information, while RAM may hold any number of different types of information.

A computer's *BIOS,* or *basic input-output system,* provides a good example of how ROM is typically used. A machine's BIOS chip stores the machine's most basic instructions that it uses to communicate with peripheral components like video hardware. Without this BIOS chip, the computer would be unable to communicate with these devices. This type of information also rarely changes, and because of its importance it needs a more secure storage medium than either RAM or disks can provide.

ROM chips are generally not replaced or upgraded, and they usually do not go bad. For most users the ROM chips will be invisible pieces in their computer with which they never interact directly. It is important to know what ROM is, though, because it stores not only the machine's BIOS but other information needed by the computer and its peripherals.

A second type of ROM has emerged recently. Various forms of *flash memory* are now available. Flash memory chips are ROM chips that can be written to under certain controlled circumstances. Once written to, these chips are still nonvolatile, setting them apart from RAM. Flash memory chips are appearing more often in modems and other components, making these components upgradable through modification of the information stored in ROM rather than through replacement of the whole component.

Summary

A computer's RAM, or main memory, acts as a workspace into which the machine can load programs and data for use. RAM plays a significant role in system performance, increasing both speed and reliability. While there are different types of memory, it is the overall amount of RAM that makes the biggest difference.

RAM stores information in binary format, with bits, represented as ones and zeroes, making up bytes, which are equivalent to one character or letter. Kilobytes and megabytes are common units of storage. Main RAM is measured in megabytes, and 16 MB should be considered the minimum for most machines, while 32 MB is preferable.

Cache RAM is extremely fast memory used for storage of vital data and program segments. It comes in smaller amounts than main RAM, usually no more than a few kilobytes. Increasingly, cache RAM is part of the CPU.

ROM is nonvolatile memory usually used to store system information, such as routines for communicating with peripheral devices like disk drives or video hardware.

Glossary Terms

basic input-output system
binary
BIOS
bit
byte
cache
cache memory
cache RAM
DIMM
dual inline memory module
flash memory
kilobyte
main memory
megabyte
memory module
multitasking
RAM
random access memory
read-only memory
ROM
SIMM
single inline memory module
slot
socket
volatile

4

Storage

Storage has become a critical issue as applications grow in size and complexity. Libraries use increasingly large amounts of data stored locally on CD-ROMs, hard drives, and servers, and patrons have come to expect fast and consistent access to electronically stored information. While two or three types of disks, and their accompanying drive types, still account for a large majority of the storage media being used, other media are evolving to meet growing storage demands.

Any discussion of storage must take into account not just the volume of storage a medium provides but the security of that medium as well. In this context, it is not so much security from theft as much as it is security from accidental loss of data that is important. Data loss can be expensive and frustrating, and proper use of computer storage can help alleviate these accidents.

Floppy Disks

Floppy disks, also called *floppies* or *diskettes,* are so ubiquitous that they are becoming the computer equivalent of scratch paper. They are the quickest and most convenient form of storage for small amounts of information.

Users often wonder why floppy disks are called floppy when they are relatively inflexible. Floppy refers to the medium inside the less flexible outer casing. This medium is a mylar film, coated with a thin layer of iron oxide that holds a magnetic charge,[1] and is essentially the same material found inside a cassette tape; but instead of a long strip, it is formed into a disk. The disk is kept in the outer case for protection, because static electricity, dust, and a number of other contaminants can damage the disk's coating and cause data loss.

All diskettes have a few things in common. The disk drive clamps onto the center of the disk and spins it, allowing the *read/write heads,* or just *heads,* in the drive to read or write information to the disk as it spins. Figure 4.1 shows a read/write head and a disk. The dark arrow indicates the direction of movement the head takes as the disk spins beneath it. This is also true of hard disks, which are discussed in the next section.

The inside of the disk's outer casing is lined with cloth, which allows the disk to spin at high speeds without generating static electricity and without any of the iron oxide being scraped off by friction from the case.

Disks must also be formatted. The disk *format* divides the disk into a series of tracks, clusters, and sectors. Tracks are concentric circles running around the disk, similar to the grooves on a phonograph record. Tracks are divided into sectors, and each individual sector usually holds 512 bytes.[2] Clusters are groups of sectors. It is also important to note that current diskettes are *double sided,* meaning they can store information on both sides of the disk. A floppy drive will have two heads, one for each side of the disk. These heads move as a pair, and always read tracks that are directly opposite each other on the two sides of the disk. These opposite tracks are referred to as a *cylinder.* Cylinders are more important in hard drives, but they exist on floppies as well. Tracks, sectors, clusters, and cylinders correspond to specific physical locations on the disk, but they are not visible to the eye if the disk is inspected.

Formatting also explains the difference between disks for IBM-compatibles and Apples. Each type of machine has different requirements for reading the data on a disk, and they format the disks accordingly. Because of this, neither type of machine can read disks from the other type without software

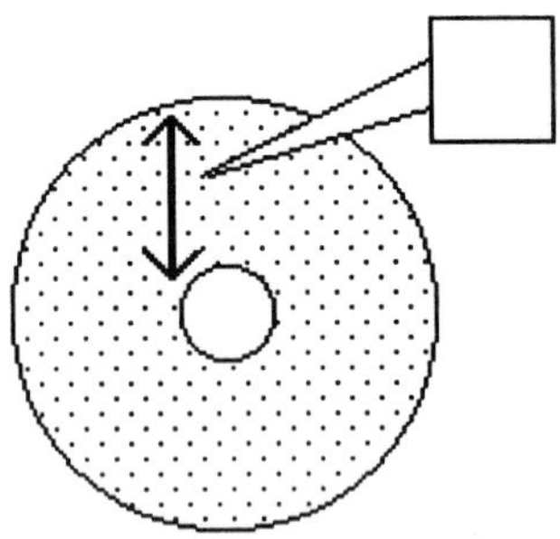

Figure 4.1. A read/write head above a disk.

enabling it to do so. In the case of the Mac OS, this software, called PC Exchange, comes bundled as part of the system. Apples will automatically detect when an IBM-formatted disk is inserted into a drive and will write data to that disk in an appropriate format. PC Exchange will also handle high-capacity media, such as PC-formatted Zip or Jaz disks.[3]

The reverse is generally not true. Data on a Mac disk will almost always need to be transferred to an IBM-formatted disk or transferred over a network in order for that data to be readable to an IBM-compatible computer. This sometimes causes great inconvenience, but it is one of the unfortunate realities of the computer marketplace. Standards developed by competing companies often lead to incompatibilities that affect the consumer.

There is an important caveat to the transfer of data between computers on floppy disks. A computer may be able to read a disk but at the same time be unable to read a file saved in the wrong *file format* on that disk. A file format is a method for encoding data that is meant to be used by a particular type of application. For example, a graphics file will be stored in a different format than a word-processing file, which will be stored in a different format than a spreadsheet file. Each file will be readable only with software designed to interpret that type of data. It is important to mention these conflicts in this context because the problem often occurs when data is exchanged on diskettes between computers. The only time a user can be certain that a file created on one computer can be read by another computer is when both computers are using identical software. For this reason, it is extremely convenient to have everyone in the same office or work area using the same type of software and same type of computer.

There are also important physical differences between disk types, which affect the way they are used. The most common types of floppy disks are 5.25″ wide and 3.5″ wide. These are often referred to simply as *five-and-a-quarters* and *three-and-a-halves.* Figure 4.2 shows a 5.25″ floppy disk. This type of diskette is obsolete and should be avoided, as it is far more vulnerable to dirt and other physical damage than a 3.5″ disk.

Figure 4.3 shows the top and bottom of a 3.5″ floppy. These disks are more secure in several ways. First, the outer casing covers almost the entire diskette. On the bottom of the 3.5″, only a small portion of the actual disk is visible, and most of this visible portion is not the disk itself but a round metal piece with several holes into which the drive clamps to spin the disk. In contrast, the 5.25″ disk is left more open in the center and also through the oval window that the heads use to read and write to the disk. The 3.5″ diskette also has a metal cover over its corresponding opening, and this cover slides open to reveal the inner disk only after the diskette is in the drive.

Figure 4.2. A 5.25″ floppy disk.

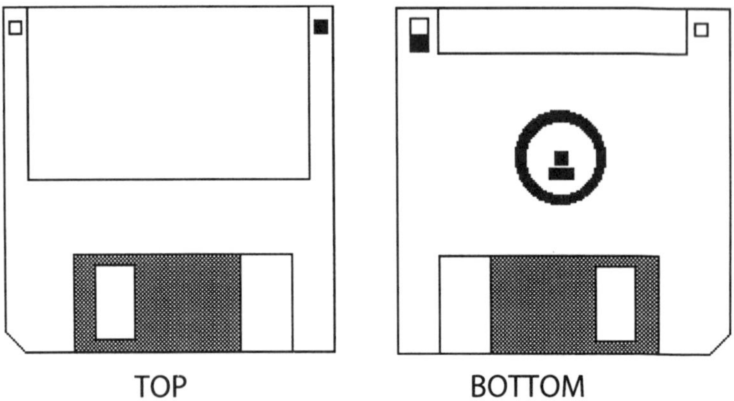

Figure 4.3. The top and bottom of a 3.5″ floppy disk.

Although the actual disks are made of roughly the same material, not all of them will hold the same amount. Floppy disks can be *double density* or *high density. Density* refers to the number of tracks the disk is capable of holding. There are actually other densities available, but they are rarely seen in normal use. Table 4.1 shows the storage capacities of common disk types.[4]

The double-density 5.25″ disks are actually no longer common, but they still turn up occasionally when data is being salvaged from an older computer. Almost all disks currently in use are high density, double-sided, 3.5″ diskettes. They are usually bought preformatted for either IBM-compatibles or Apples. If a user buys unformatted disks, or needs to reformat a used disk, *it is vital that the disk be formatted for the amount of storage for which it was intended.* Formatting a 720k floppy as a 1.4 MB diskette will not give the user a 1.4 MB diskette. It will give the user an improperly formatted diskette on which data will not be securely stored. For this reason alone, it is a good idea to buy preformatted diskettes.

Unfortunately, reading one type of floppy disk in another type of drive is not always simple. Just as there are double- and high-density diskettes, there are also double- and high-density drives. Virtually all current drives are high-density drives that have a *media sensor,* a sensor that tells the drive what type of disk has been inserted. The important warning here is that a high-density drive will read a double-density disk, but a double-density drive will not read a high-density disk.

Maintenance

Maintenance on floppy drives is virtually nonexistent. The drives are extremely inexpensive, usually under $50, and easy to replace. Because of low prices and easy installation, floppy drives are usually replaced rather than repaired. The most common problem is for the heads on the drive to go out of alignment, which means the heads no longer track properly across the surface of the disk. Heads can be realigned, but realignment will usually create more expense and require more effort than simply replacing the drive. The main reason to mention realignment at all is the effect it can have on disks. Disks formatted or written on a misaligned drive may be unreadable on

Table 4.1. Disk Capacities

Size	*Density*	*T.P.I.*	*Capacity*
3.5″	High	135	1.4 MB
3.5″	Double	135	720k
5.25″	High	96	1.2 MB
5.25″	Double	48	360k

other drives, and disks formatted and written on properly aligned drives may not be readable on the misaligned drive. Salvage the data on disks used in a misaligned drive by copying it to the hard drive, then throw the disks away once the new floppy drive is installed.

The only other maintenance that might need to be performed on a drive is to clean the heads occasionally. Cleaning kits are available, but excessive cleaning can damage drive heads. In general, only clean the heads if the computer is in a very dirty environment or if a particular drive generates disk error messages. If the cleaning fails to correct the problem, it is usually best to replace the drive.

The Future of Floppies

Any meaningful discussion of floppy disks must include some mention of the impending obsolescence of this technology. The standard 1.4 MB, 3.5″ high density is the standard for now, but the complexity of applications requires a larger amount of storage than these disks provide. They will be obsolete within a few years, and a variety of technologies are evolving to fill the gap. These technologies are discussed later in this chapter.

Hard Disk Drives

Hard disk drives, or *hard drives,* bear great similarities to floppy drives and disks, but there are also important differences that give these drives higher capacity and greater security. The principal difference between a hard drive and a floppy drive is that a hard drive contains multiple disks, and these disks are not removable. For this reason, hard drives are sometimes referred to as *fixed disks,* fixed drives, or fixed media. The individual disks in the drive are referred to as *platters,* and a hard drive will have a varying number of platters, depending on its capacity. Hard drive capacity used to be measured in megabytes but is now more frequently measured in gigabytes.

The disks in hard drives are made from a hard substance, usually an aluminum alloy. The platters are stacked close together inside the drive, with sets of read/write heads moving in between to read and write to the individual disks. The platters in most current drives are 3.5″ across, 2.5″ in many laptop machines. Drives typically contain anywhere from two to approximately a dozen read/write heads.[5]

Because hard drives hold so much more data than floppies, they are now almost always used to store the operating system and are almost always the system's *boot disk.* The boot disk is the disk the computer looks to during startup for system software necessary to run applications and perform basic functions.

Two maintenance procedures should be performed regularly on hard disk drives. The first of these is *Scandisk* or its equivalent. Scandisk is a program that scans the surface of each disk and searches for errors in the logical structure used to store data. It can also check for imperfections on the physical surface of the disk. If Scandisk finds bad clusters, it can make note of them and make sure the computer no longer uses them. This avoids both the data loss from use of a bad cluster and the expense of having to replace a hard drive that has a small number of bad clusters. Scandisk is built into MS-DOS versions 6.2 and later, and CHKDSK (short for Check Disk) is built into earlier versions. Always choose Scandisk over CHKDSK, as it has superior features.[6] Windows 95 also has a version of Scandisk built in. This capability must be added to Apple computers through the purchase of third-party software. Two of the best-known packages for this purpose are Norton Utilities and Mac Tools Pro.

In many cases, disk *defragmentation* affects hard drive performance even more than the defects detected by Scandisk. Over time, a hard drive becomes fragmented when pieces of the same file are scattered over different locations on the drive. This happens because the drive puts data in the first available spot it finds, and this may not be adjacent to other parts of a file. Defragmentation, also called *defragging* or *optimizing,* rearranges information on the drive so that the pieces of each file are next to each other. The heads then need to travel shorter distances over the platters, speeding access time. A good defragmenter, called *Defrag,* is built into MS-DOS versions 6 and later,[7] and a version is also built into Windows 95. For Apples, and IBM-compatibles running earlier versions of DOS, defragmentation software must be purchased separately.

Together, Scandisk and Defrag, or their equivalents, are among the most important maintenance procedures that a user can perform. Purchasing such software separately for Apples or older IBM-compatibles may seem inconvenient, but it is definitely worthwhile. It is less of an issue with older IBM-compatibles as they age out of use, but it is very much an issue with Apples. Purchase of this sort of maintenance software should be considered a routine part of purchasing an Apple computer.

CD-ROM

CD-ROMs, or *CDs,* short for compact disc-read only memory, are an increasingly popular storage medium because of their relatively large and secure storage capacity. CDs consist of a plastic substrate covered by a metal coating into which pits have been etched with a laser (there is also an outer protective coating). The pattern of pits and spaces is read as bits and bytes just as the magnetic charges on a floppy or hard disk.[8] The drive does this by focusing a laser, weaker than the one used to etch the disc, on the surface of the moving disc, then reading variations in the laser's reflection as it passes over the pits and highlights on the surface.

In both speed and storage capacity, CD-ROMs fall between floppy disks and hard drives. A standard CD-ROM will hold roughly 650 MB, compared to a floppy's typical 1.4 MB and a hard drive's capacity, which will most likely be measured in gigabytes. Unless the CD becomes scratched, it is extremely reliable. It might be interesting to note that, due more to historical accident than anything else, CDs are called "dis*c*s," while floppy and hard drives are said to contain "dis*k*s."

Just as the growth in the size of applications has spurred similar growth in the size of hard drives, it has also driven a switch from floppies to CD-ROMs for the installation of new software. A large suite of software, such as a suite of office applications, can easily take up thirty or more floppies. Not only is this inconvenient but, with such a large number of disks, the chance that one of them will be defective increases greatly. One defective diskette can cause the installation to fail, rendering all the other diskettes useless. Computer applications also use an increasing amount of graphic material, which requires large amounts of storage space. A CD-ROM drive should now be considered standard equipment on any new machine. Buying a new machine without a CD-ROM drive would be like having bought a computer five years ago without a floppy drive.

The main parameter to remember when comparing CD-ROM drives is the speed with which the drive spins the disc. This is measured in comparison to the speed of the original CD drives for audio disks. A 1x CD-ROM drive spins the disc at the same speed as an audio CD player, and a 4x drive, sometimes called a "quad speed," spins the disc at four times that original speed. The maximum speed available on the market is increasing frequently, and currently drives with speeds up to 32x are available. As with all computer components, more speed is almost always better. But keep in mind that the data on many CDs is optimized for a certain speed, often quad speed. Such a disc will not take full advantage of a drive's speed increase. Even keeping these caveats in mind, it is unlikely that most users will regret having a high speed CD-ROM drive.

Very little maintenance is required for CD-ROMs and their drives. Most problems occur because of dirt on either the disc itself or the laser lens in the drive. Kits are available to clean both the disc and the lens, but in practice these are rarely needed. An increasing number of CD drives clean their own lenses, and a correctly handled CD-ROM should receive no dirt or scratches. The proper way to handle a CD-ROM is to pick it up by the edges and never touch the bottom, which is the reflective surface read by the laser.

Other Storage Media

This discussion would be incomplete without some mention of other media, some of which have been with us for a while, others of which have been developed recently. Some are used in the same way floppies usually are, for day-to-day storage and transport of data, and others are intended for long-term backup.

Writable CDs

CDs are part of a larger group of storage solutions referred to as *optical media.* In this context, optical media refers to media read by a laser. These media are currently being developed in a variety of forms, and there is no one type or standard emerging as the dominant type. The most widespread is the writable CD, sometimes called *CD-R,* short for compact disc–recordable. They are also occasionally referred to as *WORM drives,* with WORM being short for Write Once Read Many. This refers to the fact that the blank discs used in these drives cannot have their data erased and written over. Once the laser etches the aluminum coating, the pits are permanent.

The prices for these drives are falling, and this, along with the familiarity of the medium to many users, has helped their popularity. CDs that can be rewritten multiple times are also available and being developed further.

Zip Disks

Developed by a company named Iomega, Zip disks combine many of the virtues of hard and floppy disks. The drive incorporates read/write heads more typical of hard drives, which can read and write more data in a smaller space than the heads from a floppy drive. The disk itself is roughly the same size as a 3.5″ floppy, with a more robust outer casing. Zip disks hold approximately 100 MB, and competitors are producing similar disks with varying storage capacities. There is currently a consensus that some form of these disks is likely to replace floppies in the near future.

Portable Hard Drives/Removable Cartridges

Hard drives do not need to be mounted inside the computer. Portable hard drives, which can be plugged into one of a computer's ports, usually the parallel port or a port on a SCSI card, are available. Some of these drives avoid fixed disks and have removable cartridges that will typically hold up to a few hundred megabytes. Both the cartridges and the drives come in various forms, and no single standard dominates.

Tape Backup Drives

Magnetic tape drives are among the most venerable forms of computer storage and are usually used for backup of archived data. The most popular of these devices uses a cassette that is larger and more robust than an audio tape, with the tape itself being a quarter inch wide. These tapes conform to the *QIC* standard, developed by the *QIC,* or Quarter Inch Committee.[9] Backup tapes are inexpensive, but they are also slow. It remains to be seen whether they will remain popular as faster forms of storage media fall in price and increase in reliability.

Digital Video Disc (DVD)

Digital video disc, also called *digital versatile disc,* but most often *DVD,* is currently emerging as a possible replacement for CD-ROMs. DVD drives and discs are an optical storage medium offering large storage capacities, anywhere from 4.7 GB to 17 GB, the ability to read traditional CD-ROMs with the DVD drive, and the capability to write to the discs.[10] DVD is currently used principally for the delivery of movies through a computer, but it also offers the possibility of very effective mass storage.[11] Despite this list of exciting capabilities, DVD has not made a rapid penetration into the market, and it has an uncertain future. Because most libraries have videotape collections and equipment for delivery of movies, DVD is probably not a necessary addition to their technology offerings. However, as DVD, or some alternative, matures, libraries may eventually want to reconsider this.

Backing Up Data

Backing up data may be the most important precaution users can take to safeguard their peace of mind. The rule of thumb for backing up data is simple: do it early and do it often. Floppy and hard disks, the dominant forms of storage, rely on magnetism and electricity to store data. This means that magnetic fields, static electricity, power failures, and defective media can cause loss of data. If the data, or a reasonably recent version of it, is available on a backup medium, such as a tape or portable hard drive, a great deal of time and expense can be saved. No form of backup storage is perfect, but the chances of both the primary storage medium and the backup medium failing are relatively small. Under no circumstances should floppy disks be considered secure, permanent storage. They are simply not robust enough.

Some users back up their entire hard drive. This has the advantage of saving the operating system, settings, drivers, application software, and data, but it has the disadvantage of being time consuming and more expensive.

As an alternative, many users only back up data. The backup of data is generally more important than the backup of application software. This method is quicker because it backs up a smaller volume of information. And if only the data file on the hard drive is corrupted, then replacement is quick and easy. If application files are corrupted also, the application can simply be reinstalled from the original installation media.

Backing up the entire hard drive has advantages, though, if there is a total failure of the drive. If the backup works flawlessly, the original system, complete with individual settings on programs, can be restored to a repaired or replacement drive. This can be easier than reinstalling everything from the operating system on up and then restoring preferred settings. Many users compromise between these methods, backing up important data and vital system files. They are willing to reinstall the rest a piece at a time if need be.

One thing can be agreed on, though. Smart users back up *at least* their important data. For example, the data in a collection development survey would be expensive, time consuming, and difficult to replace. Such vital data should absolutely be backed up, perhaps more than once. Users with vital data might want to consider backing it up twice and keeping one of the backups in a separate location as a form of disaster planning.

Summary

The growth in size and complexity of computer software has spurred the development of storage media capable of storing larger amounts of data more securely. The most common storage media are 3.5″, double-sided, high-density floppy diskettes; hard disk drives; and CD-ROMs.

Floppies are portable and convenient but are not robust enough to be considered permanent, reliable storage. This, along with their relatively small capacity, guarantees their eventual obsolescence.

Hard drives provide greater storage capacity, higher speed, and more security than floppies. However, the majority of hard drives are fixed in the computer, eliminating them as portable storage. Most computers use the hard drive as the boot disk, making maintenance of the hard drive one of the most vital actions a user can take.

CD-ROMs provide relatively large amounts of very secure storage. They are beginning to replace floppies as the medium of choice for installing new applications, and a CD-ROM drive should be included in any new computer. The major disadvantage of standard CDs and CD drives is that the media cannot be rewritten. It can only be read.

A variety of storage media are evolving to meet short- and long-term storage needs. Writable CDs and backup tapes are among the more popular, but there are a number of competing standards, and no single one is appropriate for all purposes.

Backing up important data is vital, and users may want to back up some application and system files as well.

Glossary Terms

boot disk
CD
CD-R
CD-ROM
cylinder
Defrag
defragmentation
density
digital versatile disc
digital video disc
diskette
double density
double-sided
DVD
file format
five-and-a-quarter
fixed disk
floppy
floppy disk
format
hard disk drive
hard drive
heads
high density
media sensor
optical media
optimization
platter
QIC
read-only
read/write heads
Scandisk
tape drive
three-and-a-half
tracks per inch/t.p.i.
WORM drive
writable

5

Monitors and Video Cards

Few things affect user satisfaction more than a computer's video display. A blurry monitor can make an otherwise good computer look cheap, but a high quality monitor with a good video card can not only make the display more attractive but faster as well. Speed and appearance are two of the things users notice most, and making sure that this part of a system is running well goes a long way toward keeping users satisfied.

Unfortunately, the parameters that go into making up a good display are among the most complex of any computer subsystem and depend on the features of both the monitor and video card. Neither by itself will give a good display. This is another case, though, where learning a few terms and acronyms, and demanding certain minimum specifications from any new system, will help ensure user and staff satisfaction.

Monitors

The first specification to consider in a monitor is screen size. It is important to note that this is a measurement, in inches, of the picture tube's diagonal width, not the viewable area of the screen. The plastic bezel that forms the front of the monitor's cabinet invariably obscures a portion of the tube. The three most common monitor sizes, with their listed width and the average viewable area, are listed in Table 5.1.

Both 14″ and 15″ monitors are adequate for one or two users who are working on one application at a time. But when affordable, 15″ monitors should be considered the standard. They are noticeably larger and generally do not cost significantly more. If a computer will be used regularly by groups

Table 5.1. Typical Monitor Sizes and Viewable Areas

Tube Width	*Viewable Area*
14″	~13.5″
15″	~14″
17″	~16″

of more than two or three users, consider a 17″ monitor. This will make it significantly easier for group members to see the screen. The price jump from 15″ to 17″ can be significant, though, so infrequent use of a machine by such medium-sized groups may not merit the larger display.

In addition to different screen sizes, there are also different screen types. The most desirable is a flat screen with a nonglare coating. A flat screen is significantly less curved than a normal picture tube and thus picks up glare from fewer angles. The anti-glare coating further reduces eye fatigue. Detachable filters, made of plastic or cloth, are also available and provide the same effect, but not as well. Flat, coated screens add to the cost of a monitor but can be well worth the extra money. In any area where there is significant glare from artificial light, or outside light sources, a flat screen with some sort of coating or filter should be considered. If individual users will be using a computer for more than a few minutes at a time, these features should be considered a necessity.

Screen resolution is the next consideration, and is closely related to screen size. Any screen image on a computer monitor is composed of thousands of small dots called *pixels,* which is short for *picture elements.* A monitor's resolution is expressed as the number of pixels across and down the screen. So a monitor running at a resolution of 640 × 480 has 480 horizontal rows, each containing 640 pixels. This should be the minimum resolution of any recently manufactured monitor.

Screen resolutions come in standard increments, and the next standard resolution is 800 × 600. As the number of pixels on the screen increases, the size of each object on the screen will shrink, because each object is still made up of the same number of pixels. An application window, for example, would shrink, leaving room for the user to open up a second application and view them side-by-side. The main advantage of higher resolutions is this ability to have more than one program on the screen at a time. But at extremely high resolutions, objects on the screen can become uncomfortably small.

For 14″ monitors, a resolution of 640 × 480 should be considered standard, with 800 × 600 being the standard for 15″ displays. Quality monitors of either size will be capable of displaying finer resolutions, but many users will be unhappy with the resulting size of on-screen objects.

The next standard resolution is 1024 × 768, and this is appropriate for 17″ monitors and larger. However, many users will be quite comfortable with a resolution of 800 × 600 on a 17″ screen. Larger monitors with increasingly fine resolution are available but are unnecessary for most library applications. They are principally specialty products that are useful in graphics-intensive applications, like animation or illustration. Even for multimedia CD-ROMs, or Web pages with heavy graphics, a 15″ or 17″ monitor will keep all but the most demanding users happy.

A factor closely related to screen resolution is a monitor's *refresh rate.* The image on the screen is not static but is refreshed, or redrawn, dozens of times per second. The refresh rate is a measure of how fast the monitor does this. A monitor with a refresh rate of 70 Hz (hertz, or cycles per second) redraws the picture seventy times each second. A high refresh rate will help produce a sharp image that appears stable on the screen. A low refresh rate can cause the image to flicker, especially around the edges. A flickering image is extremely uncomfortable to look at and can cause eye fatigue and headaches in a matter of moments if the flicker is severe enough. The absolute minimum refresh rate for a monitor should be 60 Hz. A refresh rate above 70 Hz is preferable, especially if users will be viewing the screen for more than a few minutes at a time.

The refresh rate will generally decline as the resolution goes up. This is not universal, but many monitors will have a slower refresh rate at 1024 × 768 than at 640 × 480, for example. A monitor's useful range of resolution ends when its refresh rate drops below 60 Hz. Most commercial monitors now have an acceptable refresh rate at both 640 × 480 and 800 × 600, but that assumption should never be made.

When the display is refreshed at most resolutions, the redrawing starts at the top and proceeds line-by-line down the screen. On an *interlaced* monitor, though, this is not true. The monitor redraws every other line, then goes back and redraws the lines it left on the first pass. This has the effect of slowing down the redrawing of the screen and should be avoided. Most monitors are noninterlaced but may switch to an interlaced mode at high resolutions. Just as a monitor's useful range of resolutions ends when its refresh rate drops below 60 Hz, its useful range will generally end when it has to go into interlaced mode.

All monitors should be capable of displaying at least 256 colors. Depending on a monitor's specifications, it will be capable of displaying anywhere from 16 to many millions of colors. With 256 or more colors, a monitor will do justice to most graphics displays. Most monitors and video cards are now capable of much larger numbers of colors, but the added capability is a hedge against obsolescence rather than a current necessity.

The final consideration with a monitor is its *dot pitch interval* (also known as *dot pitch, dpi,* or just *pitch*), which is a measure of how tightly packed the pixels in a screen image are. The more closely the pixels are spaced, the sharper the picture will appear. This measurement of a monitor's performance should not be confused with the term *dpi* as explained in the chapter on printers.

Each pixel, or dot, that makes up the screen image is composed of three smaller dots—one red, one green, and one blue. Varying intensities of these three colors are blended to form other colors. Dot pitch is a measurement, in fractions of a millimeter, of the distance between adjacent dots of the same color. So if a monitor has a dot pitch of .28, the distance between the red dots in adjacent pixels is .28 mm. Figure 5.1 shows the way in which the pitch between two adjacent pixels is measured.

What is important to remember about dot pitch is that the smaller the number, the sharper the image on the screen will appear. On 14″, 15″, and 17″ monitors, the standard increments are .39, .28, and .25, though other increments are available. A pitch of .28 should be considered the standard. It will yield a perfectly adequate picture on monitors in a reasonable price range. A pitch of .39 will often yield a picture that is noticeably fuzzier, and it should be avoided in any monitor which will be used for more than a few minutes at a time.

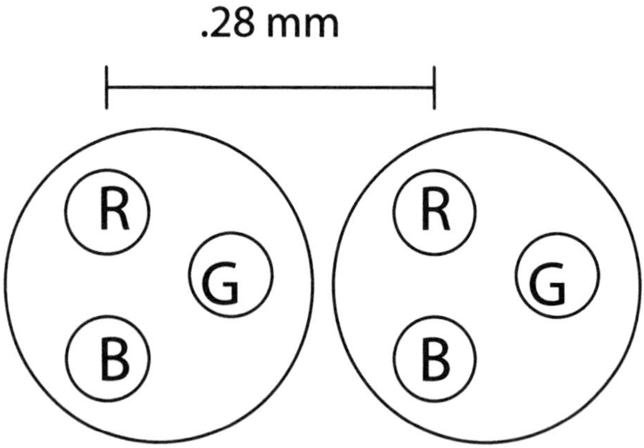

Figure 5.1. Dot pitch interval (dpi) on a color monitor.

Video Cards

Video cards, often called *video accelerators,* are circuit boards that occupy one of the computer's expansion slots, and convert the computer's data into a form the monitor can understand. These cards play a crucial role in the quality of a computer's display by providing *video RAM* (separate from the system's main memory) and video processing. While the monitor plays the larger role in the sharpness of the on-screen image, video cards have a greater effect on the speed with which pictures and animation are displayed. Users often judge a computer's performance and speed on how quickly images are displayed, so a video card's performance can strongly affect user satisfaction.

The most important specification of a video card is the amount of video RAM it has. This memory is separate from the system's main memory, but it is also measured in megabytes. Memory capacity on video cards generally ranges from 512k (half a megabyte) up to 8 MB. For any new system, 1 MB should be considered the minimum, and 2–4 MB is preferable. Older video cards with less memory than this can often be upgraded by adding additional RAM, which can improve both speed and image quality. The original documentation for your system should tell you which type of video RAM you will need if you want to add more. If the documentation is missing, a good vendor or the card's original manufacturer should also be able to supply the necessary information.

The video memory also affects a monitor's performance in other ways. To produce a certain number of colors at a given resolution—for example 65,000 colors at 800 × 600 resolution—your system will need a minimum amount of video memory, 2 MB in this case. As with the system's main memory, it is nearly impossible to have too much.

The amount of video RAM and the quality of the card also determine what refresh rate a card is capable of providing for a monitor. If a monitor is capable of a higher refresh rate than a video card will provide, the extra capability of the monitor will go to waste. Usually, though, this will not be a problem. Most video cards are capable of faster refresh rates than most of the monitors on the market. Just be certain that any system you buy is capable of displaying the number of colors you will need at the resolution and refresh rate you want. Never make assumptions.

In the past, video circuits were often built into the motherboard rather than placed on a separate expansion card. A few systems are still configured this way, and if possible, they should be avoided. It can be difficult to upgrade from the motherboard's video system to a separate card, and motherboards often allow for less expansion of video RAM.

There are a variety of graphics file formats available, ranging from simple pictures to complicated, 3-D animations and full-motion video. A graphics card with the recommended 2–4 MB of memory should have no trouble providing an adequate display of any of the formats. The differences in cost, amount of video RAM, and so on, will mainly effect the speed and sharpness with which these images are displayed. An exception to this are movies in *MPEG* format (pronounced *em-peg*). MPEG is an acronym for Motion Picture Experts Group, and files in this format provide high-quality video and audio. These movies have special hardware and software requirements, which an otherwise good video card may not necessarily meet. For the sake of future compatibility, it would be wise to buy a system that will support MPEG movies.

Summary

The specifications and performance measurements of monitors and video cards can be complex. Fortunately, a few rules of thumb and a dose of common sense is really all that is needed in evaluating a system's video performance.

For most library applications, a 15″ monitor should be considered the standard. The monitor should have a dot pitch of .28 or less, and should be capable of displaying at least 256 colors at a resolution of 800 × 600 with a refresh rate of 70 Hz or higher. An increasing majority of commercially available monitors will meet, or usually exceed, these specifications. However, not all will, and some caution is still needed.

Video cards now included in most computers also meet or exceed the specifications most libraries will require. The most important factor is to get a video card with 2–4 MB of video RAM, with 1 MB being the realistic minimum on economy systems. The card needs to be capable of a refresh rate at least as fast as the monitor's.

These specifications also apply to older systems. Any existing system that cannot be upgraded to meet these specifications should be considered a candidate for replacement. The increasingly complex nature of the graphics in many software applications and Internet sites may produce unacceptable performance on older systems.

Glossary Terms

dot pitch
dot pitch interval
dpi
interlaced
MPEG
picture element
pitch
pixel
refresh rate
screen resolution
video accelerator
video card
video RAM

6

Printers

Printers are the most important output devices used in most settings. Low-quality printouts can make even a good computer look bad, and faulty printers can take a great deal of staff time and energy. On the other hand, high-quality printouts can make a library's handouts, internal documents, and so on, look highly professional.

Printers range widely in price and quality, but as with most other things discussed in this book, they can be differentiated by looking at a relatively small number of specifications that give a reliable picture of a printer's quality. The most basic difference between printer types is the way in which they place text and graphics on the page. Three basic printer types are in common use in libraries—dot matrix, inkjet, and laser printers—and each has a different type of design and construction.

Fonts

All printers have several things in common, and the most important of these is the use of *fonts,* or *typefaces.* A font is simply a graphic style of text, and each font is actually a data file that must be present for a printer to produce that particular typeface. Word processing applications usually come with a large number of fonts of their own, and many printers have fonts built in. In addition, fonts can also be purchased separately like any other software, but this tends to be quite expensive. Fonts are usually bought individually only when a particular font is to be used for an advertising campaign, institutional logo, or similar application.

Most fonts are *scalable,* meaning they can be adjusted to a large number of sizes on the page or screen. Font size is measured in *points.* A point is 1/72 of an inch, and a 12-point font would thus produce characters 1/6 of an inch high on the printed page.[1] A font of 10 or 12 points is the norm on a standard-sized sheet of paper.

Dot Matrix Printers

Dot matrix printers are among the most rugged and reliable printers, but they also have the lowest print quality and produce the most noise. Quieter dot matrix printers, and ones with higher print quality, are available, but they are being supplanted by inkjet and laser printers, which will be covered in later sections of this chapter. Despite their declining popularity, dot matrix printers are still useful in certain settings and can provide an economical alternative to other types of printers.

Construction

The heart of a dot matrix printer is the *print head.* The print head is the part of the printer that moves over the paper and does the actual printing by placing a series of small dots on the page. These dots make up the text or images. Dot matrix printers are often called *impact printers* because these dots are placed by *pins,* small pin-like pieces of metal that *fire,* pushing the printer's ribbon against the paper to place each dot. Firing simply refers to the outward and return motion of the pin. The pins in the print head must fire thousands of times to make up the characters on a page of text. These impacts create the signature high-pitched sound of a dot matrix printer.

Dot matrix printers create letters and images by printing dots within a matrix, or grid. Figure 6.1 shows how the letter *A* might be printed. This is a simplified example. A real printer would probably print the letter *A* using more dots to add detail, but the figure expresses the basic idea behind using a matrix of dots to represent a character.

One of the main differences between models of dot matrix printers is the number of pins in the print head, usually 9 or 24. The 24-pin printers generally produce higher quality output simply because a larger number of dots can produce finer resolution. But depending on the use to which the printer is put, 9-pin models may still be quite adequate.

Some newer dot matrix printer designs incorporate technologies that reduce the amount of noise produced by the impact of the pins. In many older models, and some new ones as well, the pins are arranged in a straight line.

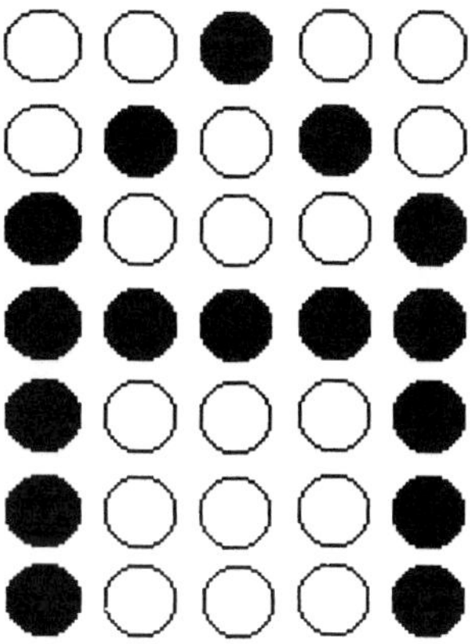

Figure 6.1. A dot matrix *A*.

The pins may all fire at once to produce, for example, a vertical line. Having multiple pins firing simultaneously increases the amount of noise given off during printing. Quieter models arrange the pins in a different pattern, often one similar to a parenthesis or crescent. In this arrangement, fewer pins pass over an area and fire at any given moment, reducing the noise. Printers with this noise-reducing feature tend to print more slowly in the quiet mode, so many of them can be switched to a faster, but noisier, mode.

The speed of a dot matrix printer is usually measured in *characters per second,* or *c.p.s.* This is a measure of how many characters of text a printer can print in a second. A printer's speed will vary depending on the quality of the individual printer and on the mode the printer is in, if the printer is capable of more than one. Many dot matrix printers allow the user to choose between *draft mode* and *near letter quality,* or *NLQ* for short. Draft mode is often quite fast, with speeds exceeding 70 or 80 c.p.s., but the output is of inferior quality. NLQ mode, on the other hand, can slow a printer down under 30 c.p.s., but produces better looking output and is capable of using a greater number of fonts. Many dot matrix printers will not print graphics in draft mode.

Use

Dot matrix printers are becoming steadily less popular, but they can still play a useful role. In any environment when high output is needed without extremely high quality, a dot matrix printer can shine. Many libraries use them on OPAC terminals when only catalog entries or bibliographic citations are being printed. These printouts are generally only for the patron's temporary use in finding the material, not for presentation to anyone else. Similarly, staff terminals used to print out temporary cataloging slips, overdue notices, and so on, are good candidates to receive dot matrix printers.

The one clear advantage dot matrix printers maintain over other types is *per-page cost.* The fabric ribbons used in these printers are generally inexpensive and will print a very large number of pages. Per-page cost is often figured by taking the estimated number of pages a ribbon will print effectively, calculating how much that amount of paper will cost, then dividing the cost of the ribbon by the number of pages. So for example, a ribbon that costs $10 and will print a thousand pages, with the paper costing 1 cent per page, would give a total per-page cost of 2 cents per page.

Per-page cost is also referred to as the cost of *consumables.* A consumable is anything meant to be used up and thrown away, in this case, paper and printer ribbon. The cost of consumables is climbing, as anyone who has to budget for supplies can attest. In an austere budgetary atmosphere where pennies in savings are important, dot matrix printers can still make a contribution.

Driver Software

Dot matrix printers are perhaps the only major peripheral left that does not require specific driver software. Both Windows 95 and 3.1 allow the machine to assume it is talking to a *generic/text-only printer.* This setting allows the printing of lines of plain text with no graphics and little page formatting. For this reason, dot matrix printers are sometimes called *line printers,* because the printer simply prints one line of plain text at a time. More advanced printers, such as lasers and inkjets, set up the entire page before printing any of it.

Some of the advanced features of dot matrix printers require driver software, but even with this software, dot matrix printers tend to have a smaller number of capabilities than inkjet and laser printers.

Maintenance

The most common maintenance procedure with dot matrix printers is the clearing of paper jams. This is routine and straightforward, and along with ribbon replacement is virtually all the maintenance required. Dot matrix printers tend to be rugged, long-lived, and reliable.

Inkjet Printers

Inkjet printers, also called inkjets or *bubblejets,* are generally a step up in quality from dot matrix printers, and in many settings they perform the workaday, generic printer tasks that dot matrix printers used to perform. A high-quality inkjet can approach the high quality of a laser printer, but at a lower price.

Construction

Inkjet printers work by spraying bubbles of ink onto the paper. The individual bubbles are very small in size, giving even a modest inkjet printer much better resolution than a dot matrix model. This resolution is measured in *dots per inch,* or *dpi.* The term *dpi,* when used in this context, is very different than when it is used to indicate a monitor's dot pitch. It refers to the number of individual dots of ink a printer can place on the paper per inch.

The quality of a printer's output increases as the number of dots-per-inch capability increases. As a general rule, 360 dpi is the minimum resolution needed for good quality output, and virtually any inkjet printer from a major manufacturer will be capable of this. Some printers will also have a draft mode, with resolution as low as 180 dpi, to conserve ink when printing documents that do not need a finished appearance.

The resolution of an inkjet printer can be considerably greater than the minimum 360 dpi. Reasonably priced bubblejets now have resolutions over 1,000 dpi, yielding extremely high image quality, especially for graphics. Some of these printers are capable of producing near-photographic quality for pictures. This increased performance comes at a higher price, though. Inkjet cartridges range widely in price, and while some are recyclable, i.e., they can be refilled with ink, many are not. It would be wise to find this out from the vendor or manufacturer before purchase, especially if a library wants to settle on a particular type of printer as its standard. This kind of standardization can save money by allowing the library to buy cartridges in bulk, but only if the cartridges are priced reasonably.

A printer's resolution is sometimes measured with a split number, such as 600 × 300 dpi. This means that more dots per inch are printed vertically than horizontally. A 600 × 600 dpi printer will produce better output than a 300 × 300 dpi printer, and a printer capable of 600 × 300 dpi will produce output with quality between the first two.

Inkjets have a print head just as dot matrix printers do, but the construction is fundamentally different. The print head is generally built into the *ink cartridge,* also called the *toner cartridge,* which holds the printer's ink. The print head itself will contain *nozzles* through which the ink is sprayed. The print head and nozzles are thus disposable, cutting down on the maintenance needed if they were to clog with long-term use.

Color inkjet printers require a cartridge with colored ink. Just as monitors produce colors by blending a small number of primary colors—usually red, green, and blue—a color inkjet cartridge will usually contain three colors of ink which are blended to form other colors on the page. As is the case in so many areas of computing, another warning is merited here. It is best to have a color printer that holds two separate cartridges, one with black toner, and the other a three-color cartridge. Some color inkjets come with a single cartridge containing black toner plus the three colors to be blended. In most library and office settings, though, most of the printing will be text only, which usually is in black and white. For this reason, the black toner will usually run out before the colored toner does, and when the user discards the cartridge, they are also discarding unused color toner. It is more economical to have two different cartridges, and replace them individually as they run dry. Other color printers only come with a three-color toner cartridge and attempt to create black text by blending the three basic colors. This often results in poor quality black text, and these printers should be avoided unless they will be used almost exclusively for color printing.

Some high-quality color inkjets containing color cartridges that blend more than three basic colors are now appearing on the market. The color output from these printers can be extremely good, but for many library applications they are more than is needed. In most library settings, a good *monochrome* printer, one that prints in only one color, such as black, will suffice for most or all user and staff needs.

Use

As indicated above, inkjet printers are becoming the workaday printers for many purposes. However, these printers do have their shortcomings, and they are not ideal for every situation.

Environmental factors, such as heat and humidity, have a greater effect on inkjet printers than on dot matrix or laser printers. For this reason, the output of these printers is the least consistent of the three main types. The output from almost all of the major brands of inkjet printers will be sufficient for intraoffice use, or user printouts of information from CD-ROMs or the Internet. But the quality may not be good enough for finished reports or any document that needs to look truly professional. That type of work is still largely the domain of laser printers, although the quality gap is continuing to narrow.

Inkjet printers may not always be the best choice for user terminals where large amounts of printing are done. For example, the printouts done at OPAC terminals generally do not need to be of particularly high quality. They are generally for temporary use by patrons who are searching for material in the stacks or online. The same is true of data printouts from CD-ROMs and online services. These printouts are generally used for reference purposes, not inclusion in a final report. In all of these settings, a quiet dot matrix printer can be an economical alternative to an inkjet. However, the price gap is shrinking.

But having said all of this, inkjets still occupy a large and growing niche. For World Wide Web stations, where the printing includes both text and graphics, inkjets provide a good balance between graphics capabilities, text capabilities, cost, and speed. They also do a better job of printing charts, spreadsheets, and basic graphics in word processed documents than a dot matrix, without the cost of a laser. Inkjets also tend to produce a relatively small amount of noise.

Driver Software

Inkjets, because they are more versatile than dot matrix printers, almost always require driver software. This software often comes bundled with additional fonts, desktop publishing programs, or graphics applications. Color inkjets often come with better software which is meant to show off the printer's color capabilities. Much of this software is aimed at the home user, though, and may have limited usefulness for most libraries.

Maintenance

Inkjet printers require very little maintenance. The cartridge and print head are replaced when the toner runs out, and almost no maintenance is required during the life of the print head. The individual nozzles can become clogged, but clearing them is usually a minor procedure. Many inkjets have a cleaning feature that clears the nozzles by blowing a small amount of air through them. The printer's documentation should include instructions on how to do this.

There is one final maintenance issue with inkjet printers. The type of paper, and the way it is fed into the machine, can effect a printer's ability to feed the paper properly and the quality of the printer's output. For most purposes, plain copier paper will suffice. Remember, though, that copier paper usually does not have the same texture on both sides. It is made to be printed on one side rather than the other. If you use copier paper in an inkjet printer, find out which side this is, and make sure that is the side that faces the print head as the paper passes through the printer. Some printer manuals tell you which way the paper faces, but with other printers the user has to figure that out. Unfortunately, too many methods for feeding paper exist for there to be a general rule of thumb as to how to place the paper in the input tray.

There are also special papers on the market that are designed to give enhanced output in inkjet printers. Sometimes it is simply a higher-quality paper, but some of these papers are coated with chemicals that help them to hold a better image. In general, this is more of an issue with color printing. The best idea in evaluating specialty papers is to read the printer's manual and try a small sample of the special paper. Sometimes it will give much better results than plain paper, and sometimes it will not, but it will almost always increase per-page cost, sometimes dramatically. Give these special papers a try, but buyer beware.

Laser Printers

Laser printers offer the highest quality, greatest speed, and lowest noise level of the three main types of printers, but at a higher cost. Their use is generally restricted to settings where professional quality output is required, or office settings where a large number of pages must be produced in a networked environment.

Construction

Laser printers operate in a fundamentally different way than dot matrix and inkjet printers. They are far more complicated and operate, in fact, in a manner closer to that of a photocopier than that of inkjets or dot matrix printers. The important fact to remember about a laser printer is that it uses a combination of heat and the electrical charges produced by the laser to fuse the toner permanently with the paper.[2]

The sharpness and precision of the laser provide the quality output associated with these printers. While laser printers can be affected by environmental variables, this is much less the case than with inkjet printers. They also do not suffer from as many paper feed problems as less expensive printers. For these reasons, laser printers provide the best output of the three main types, along with the most reliable performance.

Color laser printers are also available and work on the same basic principles as a black-and-white model. A high-quality color laser printer can produce pictures of photographic quality. But this increase in quality comes at a steep increase in price. A color laser printer will generally cost $3,000 or more. The prices are coming down, but these printers are still too expensive for most library settings, and their capabilities go beyond what most libraries would realistically need.

Use

As mentioned above, laser printers serve best as networked office printers when large numbers of pages need to be printed reliably. The speed and versatility of a laser printer are well suited to that environment. Not all printers can be networked to produce the output from multiple machines, but laser printers are more likely than other types to have this ability.

The prohibitive cost of color laser printers often leaves open a niche for color inkjet printers in a setting where the laser would otherwise be the printer of choice. It is not at all unusual for a laser to be used for text-only printing, or printing where black-and-white graphics are acceptable, with a color inkjet also present for print jobs requiring color. Because color inkjets are less likely to be networkable, that printer will often be attached to the computer belonging to the person who uses it most. This may seem like an inefficient arrangement, but it will almost always be less expensive than buying a color laser printer.

Driver Software

Laser printers generally come with the same type of drivers as inkjet printers. But because they are intended for office use, laser printers often come with a smaller variety of bundled software. Most offices have no use for the consumer software included with many color, or even black-and-white, inkjets.

Maintenance

As with inkjets, the main maintenance required for laser printers is replacement of the toner cartridge. However, when something does go wrong with a laser, repairs are generally more complicated than with dot matrix or inkjet printers. For this reason, many institutions that normally perform their own service will purchase maintenance and repair contracts on their networked laser printers. Only very large institutions are likely to have in-house expertise to perform those repairs.

Other Printer Types

There are other types of printers than the three discussed so far, but they have very limited use in most library settings. *Plotters* are printers which use a *pen,* or a series of pens, to draw an image on the page. The pen can place a continuous line on the page, as opposed to the small dots produced by other types of printers. Plotters are usually used to produce design or architectural drawings, and their library use would largely be restricted to facilities serving patrons with those needs.

There are also a number of printer types that use various forms of wax or plastic instead of toner. These printers are capable of extremely high-quality output, but their use is largely confined to publishing or other settings where detailed, photographic-quality output is required.

Printer Purchase

Choosing a printer to purchase can be extremely confusing. Models come and go quickly, and an individual company may offer a number of very similar models with only minor differences in quality and number of features. This is more true of printers than of other peripherals.

Reading reviews of printers can be an excellent way to sort through the myriad models on the market at any one time. The magazines included in this book's bibliography are good sources, but it can be difficult to check multiple issues of each magazine to find a review of the appropriate printer. Fortunately, Yahoo, on the World Wide Web (www.yahoo.com), compiles reviews from a number of magazines and organizes them by brand and model. Look under the URL "www.yahoo.com/_Computers_and_Internet/Hardware/Peripherals/Printers" for this information.

Summary

There are three main types of printers suitable for most library use: dot matrix, inkjet, and laser.

Dot matrix printers are sturdy, inexpensive performers that can be good printers in any environment where text-only printouts are needed. Their output is unspectacular, but so are their price and maintenance needs.

Inkjets are rapidly becoming the workaday printers in settings such as World Wide Web stations, where both text and graphics need to be printed. The print quality of a good inkjet printer can approach that of a laser, but an inkjet's output will usually be less reliable. Color inkjets are becoming less expensive and can provide high-quality graphics without the expense of a color laser.

Laser printers are best suited to a networked environment where a large number of pages must be printed quickly, or one in which professional quality printing is required. They produce sharp, reliable output but may be more expensive to maintain than dot matrix or inkjet printers.

Glossary Terms

bubblejet
characters per second
consumables
c.p.s.
dot matrix printer
dots per inch
dpi
draft mode
fanfold paper
fire
font
generic/text-only printer
impact printer
ink cartridge
inkjet
inkjet printer
line printer
monochrome
near letter quality
NLQ
nozzle
per-page cost
pin
pen
plotter
point
print head
scalable
toner cartridge
typeface

7

Sound Cards

Most computers have a built-in speaker, but the quality of sound these speakers produce is generally very poor. As a result, *sound cards* are among the most common expansion cards. They allow a computer to play back, and often record, sound with much greater fidelity than would be possible with only the internal speaker.

Sound is also one of the areas in which there has traditionally been a gap between Apples and IBM-compatibles. Apples come with built-in sound capabilities, but not all IBM-compatibles do. Because the IBM-compatibles are used in office environments more frequently than Apples, they often come without sound cards simply because office users tend to have much less need for them. The built-in speaker is sufficient to produce warning beeps and the other basic sounds a computer needs to make. The increasing prominence of multimedia applications has changed this, though. Most computers come with sound cards now, although a few machines, typically ones in a low price range or ones intended only for business use, still come without one.

Libraries increasingly want their patrons to be able to use multimedia CD-ROMs and Web sites, and this has made sound cards a near necessity. Awareness of how sound cards are built and used is an important part of deciding which machine is best for a library's needs. Their prices have come down, so that an expansion that once seemed like a luxury or a toy is now an integral part of computing.

All types of sound cards perform the same basic function. They convert the digital signals from the computer into the analog sound which we hear. If they are used to record sound, they do the opposite, taking the analog sound and converting it to a digital pattern usable by the computer. Sound cards perform this function in several ways and with differing levels of quality.

Construction

Most sound cards have the same basic construction. They come as expansion cards, even if they are included in the initial purchase, and typically use a 16-bit ISA slot. Once they are in the computer, they provide a number of ports on the back of the machine. The number and type of port will vary, but there are four types that are typical of most sound cards.

All cards need to have a jack for speakers or headphones. *Jack* is another name for a port or socket into which the user can plug speakers, a microphone, or some other accessory. The speaker or headphone jacks are typically the only ones used in libraries. Sometimes there will be a single jack for either headphones or speakers, and sometimes there will be a separate jack for each.

Most cards also have a jack for a microphone, allowing the recording of sound. Older cards may only be capable of playing back prerecorded sounds, but those cards are rarely seen now.

A port for a joystick or game controller is also very common on sound cards. Both devices are used instead of the keyboard or mouse for playing games. They often come in the shape of an airplane's joystick, hence the name. While these devices may seem to have little to do with sound, the presence of their port on a sound card represents a holdover from the time when sound cards were mainly accessories for games.

A *MIDI* port is also found on most recent sound cards. MIDI stands for *M*usical *I*nstrument *D*igital *I*nterface, and it provides an interface between the computer and an electronic instrument, such as a synthesizer. Music libraries occasionally use a sound card's MIDI features, but these devices are quite specialized, and most libraries will have little use for the MIDI port.

Sound cards, in addition to their external ports, often have an internal connector for a CD-ROM drive. In fact, sound cards, speakers, and a CD-ROM drive are often sold as a bundled *multimedia kit.* CD-ROMs are currently the best medium for the large sound and video files required by multimedia applications. Because of this, sound cards and CD-ROM drives share a large number of tasks. Many CD-ROM drives require an expansion card as an interface with the computer, and putting that interface on the sound card leaves one more expansion slot free.

Unfortunately, not all sound cards are compatible with all CD-ROM drives. If a machine has either a CD-ROM drive or a sound card, and the user wants to add the other, the newer device must be compatible with the one already in the machine. This alone is good reason to either buy a full multimedia kit, which should have compatible components, or to buy computers with the components already installed and configured properly.[1]

A standard called *MPC,* short for Multimedia Computer, is often mentioned in this context. MPC dictates the minimum configuration considered sufficient for a computer to run multimedia applications. This is a *minimum* standard, and it is always a good idea to purchase a computer that exceeds it. The most recent specifications for MPC 3 are listed below.[2] Realistically, these specifications are no longer adequate. The CPU speed, the amount of RAM, and the size of the hard drive should *at least* be tripled, and Windows 95 should be considered the standard operating system.

- 75 MHz Pentium CPU
- 8 MB of RAM
- 540 MB hard disk drive
- High-density, 3.5″ floppy drive
- Quad-speed CD-ROM drive
- 640 × 480 video resolution
- 64,000 colors in display
- Serial, parallel, MIDI, and game ports
- Microsoft Windows 3.1

Notice that these specifications apply to IBM-compatibles. Because newer Apples usually have multimedia capabilities built in, this standard is not really needed for them.

Types of Sound Cards

All sound cards perform the same basic functions, but they do this in slightly different ways depending on the model of sound card and on what quality sound is desired.

There is no one standard governing sound cards as there is for certain types of hard drives or diskettes, for example. But there are a pair of de facto standards, Soundblaster and AdLib, which most manufacturers adhere to. Soundblaster and AdLib are brand names as well. The popularity of the brands has given rise to the de facto standard. Because they are the de facto standards, most multimedia software will work with both types, and cards manufactured by other companies usually will also follow one standard or the other. Older pieces of multimedia software often require users to specify which type of card they have during installation. Multimedia programs written specifically for Windows 95 have largely done away with this.[3] Sound card compatibility is generally not an issue with Apples, since this capability is built in at the time of manufacture, and Apple-compatible software will be designed accordingly.

Within the broad Soundblaster and AdLib standards, the main difference between sound cards is in the way they sample sound. *Sampling* is the process by which the cards convert normal sound to digital signals. Rather than recording continuously like conventional recorders, a sound card takes a sample of a sound a certain number of times per second, and each sample is represented digitally in much the same way a picture is represented in a graphics file. The most common sound cards use either 8- or 16-bit sampling, and you will hear reference to *8-bit sound* or *16-bit sound.* This does not refer to the bus width of the slot into which the card is inserted but rather to the amount of memory used to represent each sample. Just as higher resolution yields a better picture, 16-bit sound cards can record sound with greater fidelity than 8-bit cards. If simple speech is being recorded, 8-bit is probably adequate, and 16-bit sound is better for music.[4] Some sound cards are also capable of 32-bit sound, but this is mainly of interest to audio enthusiasts or users actually composing music on their computer.[5] While sampling is directly applicable to recording, it also serves as a general indication of a sound card's overall quality.

A more direct measurement of sound output quality is the card's *frequency response,* or the range of sound it can play back to the user. A larger range represents a better card, and a typical range would go from 30 Hz to 20 KHz.[6] Most current cards will play back sound in stereo, but this is something that should be verified before purchase.

Use and Accessories

Use

The use of sound cards is straightforward. Once they are installed, with speakers or headphones connected, they generally operate automatically. If a program has sound capabilities, it will almost always play the sound without any prompting from the user. Multimedia CD-ROMs are the applications most likely to incorporate sound, but an increasing number of Internet sites also use sound now. In addition, it is no longer unusual for tutorials that come with hardware or software to incorporate audio or video clips. As all of these uses of sound become more common, a sound card looks more and more like a necessity.

Recording sounds on a computer may not seem like a promising activity for most librarians, but this may be changing also. Bibliographic instruction modules and oral histories can be recorded digitally with a computer, and this can hold strong appeal for patrons used to interactive media. Users need to be aware, though, that audio files take up very large amounts of disk space. As an example, 16-bit recordings can require over 10 MB of disk space per minute, with 8-bit recordings taking considerably less.[7] Sound can be valuable, but it also should be used judiciously. Because they take so much storage space, sound files are also good candidates to be stored on recordable CD-ROMs.

Voice recognition represents one of the most promising sound applications for computers. Voice recognition uses a sound card and microphone to convert sound not just into a digitized recording but into commands or text the computer can understand. This has great potential for patrons and staff members with disabilities. Voice recognition can provide the ability to dictate text into a word processor or allow the user to give the computer orders to perform basic functions like opening or closing an application. Voice recognition programs can take large amounts of memory and disk space, and are very imperfect, but they hold a great deal of promise.[8] The opposite of voice recognition is *voice synthesis.* This allows a computer to produce sounds based on text, which can be a great aid to visually impaired patrons. This software is in a state of development similar to voice recognition.

Accessories

Accessories are often needed for the user to take advantage of all of a sound card's features. Most multimedia kits include both a microphone and speakers, which are the most common accessories. But some computers do not include speakers even if they have a sound card.

Speakers themselves fall into two basic categories, *passive* and *active.* Passive speakers do not amplify the sound from the sound card, relying instead on the card's built-in amplifier. Active speakers, in contrast, have internal amplifiers driven by either batteries or a separate power supply. They can significantly boost the volume of sound from the computer. The output of a sound card is measured in watts, and inexpensive sound cards may have a very low output level, under 5 watts. The output from more expensive cards can range into dozens of watts, which is comparable to some stereo systems. For voice and simple music in a library setting, the unamplified sound from the sound card will probably be sufficient.

The presence of speakers may seem at odds with the idea of the library as a place for quiet study. In many settings, headphones may be a better choice. However, headphones have their limits. Even with a *splitter,* a small plug-in accessory designed to allow two sets of headphones to plug into the same jack, a very limited number of headphones can be used at one time. For even small groups using a computer together, speakers are more practical, and the noise may be an unavoidable effect of using multimedia.

Microphones are the next most common accessory. They vary greatly in quality, but the simple microphones included with most sound cards are adequate for recording voices.

There is one sound accessory, the *subwoofer,* that most libraries will definitely want to avoid. Subwoofers are amplifiers meant to boost the bass put out by most speakers and sound cards. Simply put, subwoofers produce the kind of bass the user can feel rather than necessarily hear. They can be extremely distracting, and are mainly of interest to audiophiles or game players who like loud sound effects.

Even without an accessory like a subwoofer, remember that computers with sound cards and a CD-ROM drive are capable of playing music CDs as well as ones containing only data. This may be a useful capability for a library that has a music collection. A multimedia computer can also be considered a small-scale listening suite.

Summary

Sound cards have gone from being a luxury or toy to being a near necessity. This has been driven by a surge of multimedia applications and Web sites, and is likely to continue. It is best to buy a computer with a sound card already installed and configured, but they can be added after purchase.

Cards capable of 8- or 16-bit sound are adequate for almost any library application, and modest speakers or headphones will also be suitable for the relatively low sound output most libraries need from these cards.

Voice recognition and voice synthesis software are currently in a relatively crude state of development but can still offer help for persons with disabilities.

Glossary Terms

8-bit sound
16-bit sound
active speakers
frequency response
jack
MIDI
multimedia kit
passive speakers
sampling
sound card
splitter
subwoofer
voice recognition
voice synthesis

8

Optical Scanners

An *optical scanner,* or *scanner* for short, is another item that once seemed like a luxury or toy but now has a number of mainstream applications. The various types of scanners perform the same basic function: they take images or text on a page and encode it in digital form for use with a computer. In a certain sense, they work much like a photocopier, but rather than creating static images on paper, they create dynamic files that can be manipulated or altered in a number of ways. While scanners are not as ubiquitous as sound cards, they still possess great potential for certain information-related tasks.

Scanner Types and Construction

Scanners connect to a computer in one of several ways. They may require an expansion card to act as an interface between the computer and scanner. This will most often be a SCSI card. Other scanners connect to the parallel port. In this case the printer and scanner need to share a single parallel port unless the user installs a second one. Sharing a single port will usually not be problematic.

As with virtually all other peripherals, scanner installation is becoming easier with the advent of Plug and Play. The interface cards themselves are usually straightforward SCSI adapters, and with parallel port scanners it is usually just a matter of installing driver software. Neither should cause great difficulty.

Hand-held Scanners

Hand-held scanners are the smallest scanners and, at one point, were the cheapest. They are the least effective type, and because of declining prices for flatbed and sheet-fed scanners, hand-helds are fading from prominence. They remain in the field mainly as legacy equipment.

A hand-held scanner is small enough to be held in the user's hand and works by being dragged across the part of an image the user wishes to scan. Hand-held scanners can scan only part of the page at a time, which is one of their major disadvantages. To scan an entire page, the user must slowly scan it section by section. A good hand-held scanner will attempt to correct for uneven hand movement on the part of the user, and its software will combine multiple scans into a single page.[1] Despite these attempts to overcome its shortcomings, a hand-held scanner will not easily produce the image quality of other scanner types.

Flatbed Scanners

Flatbed scanners are similar to copy machines in both appearance and action. With a flatbed scanner, the object being scanned lays flat on the scanner just as a document is set on a photocopier. A light inside the scanner illuminates the page, and the *scan head* moves past the document, creating the image.[2] The scan head is, in effect, the mirror image of a print head. Instead of placing the image on the page as a series of dots, the scan head reads the reflected light from the page, converting the normal image to a series of digitized dots for storage.

One of the biggest advantages of a flatbed scanner is its ability to scan a three-dimensional object, like a book, with relative ease. A book can be placed on the bed of a scanner just as it can be placed on a photocopier machine. This is not possible with sheet-fed scanners.

This flexibility comes with at least one disadvantage, though. Flatbed scanners use mirrors to reflect the image through a lens, focusing the light onto elements that sense the brightness of the image. No matter how high the quality of a mirror, it will introduce some small amount of distortion into the image. Because of this, the images from flatbed scanners may not be as faithful as images produced by sheet-fed scanners.[3]

Sheet-fed Scanners

Sheet-fed scanners generally fall between small hand-held scanners and the larger flatbed models in size, and they operate by pulling sheets of paper through one at a time. As mentioned above, sheet-fed scanners can produce higher quality images than flatbed scanners because they do not have the distortion-inducing mirrors on which the flatbed models depend. Their major disadvantage is that sheets of paper must be detached from any binding, notebook, or other holding device, before they can be scanned, and they often must be fed to the scanner individually by hand. It is impossible to scan a book with a sheet-fed scanner without destroying the book in the process. This limitation restricts the use of sheet-fed scanners.

In addition to their higher image quality, sheet-fed scanners also take up a much smaller amount of desk space than flatbed scanners. Sheet-fed scanners are even available as an integral part of some keyboards.

Because of their small size and the ease with which they handle single sheets of paper, sheet-fed scanners are especially well suited to be used as virtual copy machines in an office setting. Scanning and printing can be done without the loss of much desk space, and without a trip to the copier.

Other Scanner Types

There are a variety of specialty scanners available, but they have less use in libraries than the types outlined above. Because so many libraries also house media centers, it is worth mentioning secondary types of scanners.

Slide and film scanners are designed to scan standard photographic slides or film, and some can scan negatives. These scanners generally produce very high-quality images, but the increased quality comes at a sharply increased price. These scanners are useful mainly to users with specialized graphics needs. For users who scan slides only occasionally, a *transparent media adapter* may suffice. These adapters fit over the bed of a normal flatbed scanner, and they light the scanned object from the back. The backlighting acts like a slide projector, allowing the slide's image to be seen by the scanner. These adapters are not available for all flatbed scanners, but, when they are, they can be an economical way to begin scanning transparent media such as slides.

In business settings, it is becoming more common to see scanners that are actually a number of devices combined into one: a scanner, printer, fax machine, and photocopier. These office machines carry a steep price tag, but they save desk space and may be cheaper than the individual components bought separately.

There are also extremely specialized scanners meant to scan specific media such as business cards. These scanners are generally very compact and may come with special software to organize the scanned information.

Software

Scanners require driver software just as other peripherals do, but they also require software to interpret the scanned images and make them available to the computer. There are two basic types of software available to do this.

The first type is used separately from any other graphics software on the computer. It allows the user to scan and store an image and may provide basic image editing tools such as cropping and size adjustment. But for significant manipulation of the image, the user has to open the scanned file in a separate graphics program.

The other type of software, referred to as *TWAIN* software, runs in the background and allows the user to scan an image from within a graphics application without having to run separate scanning software. The TWAIN software works with the graphics program so that scanning features will usually appear on one of the program's menus as if built-in. With TWAIN software, scanning is only a little more difficult than opening a file already on the hard drive.

The term TWAIN has an uncertain etymology. Some quip that it is an acronym for "*T*echnology *W*ithout *A*n *I*nteresting *N*ame," and others claim the name originates from the fact that the software works "twain" the scanner and the computer (or other software). In either case, it is pronounced the same as Mark Twain's last name.

Scanner Use

The increase in scanner popularity mirrors the rise in useful applications that incorporate them. The bread-and-butter use of scanners is typically the digitization of images, but there are also other potential uses which, though closely related, bear different considerations.

Image Scanning

The scanning of images is probably most closely associated with optical scanners. Much of this has come about as a result of the rush to add graphics to Web pages. Photographs, background images, charts, and other images can add life to a Web site. While some people feel that this has gone too far, causing the creation of bloated, slow-loading Web pages, graphics online are here to stay.

Web sites are not the only resource that can benefit from the addition of scanned images. An increasing number of libraries are providing users with electronic bibliographic instruction sessions, and these sessions can often be enhanced by the addition of images. Virtual library tours, incorporating scanned photos of the building, can be offered to users, as can maps of the building, diagrams and charts outlining the research process, and any number of other types of graphic information.

Another simple, but potentially impressive, use of scanned images is to incorporate them into a library's paper handouts. An institution's seal, a small picture of the building, or any number of graphic elements can be used. Any recent word processor or desktop publishing package should allow the easy insertion of a scanned image.

No discussion of image scanning can end without at least mentioning copyright. Virtually any image on a flat page can be scanned with little technical difficulty, but that does not mean there is no legal difficulty. Libraries frequently research and establish copyright policies that are posted near photocopiers, and the same should be done with scanners. In addition, libraries should take caution when scanning seemingly free images for incorporation into Web pages.

Document Preservation

Anyone working in a library realizes that a growing number of paper documents are falling apart. These documents range from manuscripts and rare books to more recent books with acidic pages. Careful handling can only prevent so much damage, and a document in the final stages of deterioration is a good candidate to be scanned.

It is vital to remember, though, that *scanning a fragile document may destroy it.* Scanners and their scan heads use varying levels of light, and give off various levels of heat, during the scanning process. Light and heat can damage any document, and many documents preserved by optical scanning are fragile to begin with. Because of this, scanners to be used for document preservation must be chosen carefully. Before purchasing a scanner for this purpose, consult an up-to-date preservation manual.

Many institutions only use scanning as a last-ditch effort to save a nearly destroyed piece. Any scanning done for the sake of document preservation should be done in consultation with a preservation specialist. Leaping into this sort of work without preparation invites unhappy outcomes.

If it is done safely, scanning of rare items has additional advantages. The first is that use of a facsimile can reduce handling of the original. If a user simply needs to read the text, for example, there may be no need to use the original. Unless there are physical elements, such as erasures or extremely fine details, the scanned copy may be every bit as useful as the original.

The other advantage of scanning rare items is that they can be made available, potentially, to the entire world. Electronic transmission of digitized pieces can allow scholars to examine high-quality copies of works that they otherwise might never see at all.

Optical Character Recognition

One of the more promising uses of scanners is for *Optical Character Recognition,* or *OCR.* OCR software uses a compatible scanner to scan documents as editable text. During normal (non-OCR) scanning, the scanner works much like a photocopier. The image it creates is simply a picture of the document that was scanned, and while a graphics program can be used to alter the image, the text in the scanned image cannot be edited with a word processor. OCR changes that, making a scanned document editable. Instead of the digital equivalent of a photocopy, the user has the same sort of text document they would have if they had taken the time to type it in. OCR provides a powerful tool for document conversion and is, in fact, being used to bring formerly print databases online. But there are limits to OCR that the user should understand.

First, OCR has an error rate higher than users have become accustomed to with other types of applications. For example, a good spreadsheet will make virtually no errors in mathematical calculations, and a word processor will virtually never make a mistake in spell checking. This level of reliability is not available with OCR software. Even the best OCR software will have trouble reading unusual fonts, and will have difficulty working around graphic elements on the page. Plan on OCR software making an error at least every few hundred characters. For documents where accuracy is vital, this may be an unacceptable level of error, and under any circumstances, documents scanned by OCR software will have to be proofread. OCR does not preserve document formatting, either. Margins, fonts, italics, columns, will all be lost and must be reinserted by the user.

Users of OCR also need to remember that OCR is optical character recognition, and it will not recognize graphics. Graphic elements of a document will either be omitted entirely or interpreted as nonsense text.

Even with these limitations, OCR can be a powerful tool. If the documents being scanned are solely or primarily text, printed in a reasonably normal font and in English with no special characters, OCR scanning can save a user large amounts of time, even taking proofreading and reformatting into account. In addition, OCR can save a great deal of disk space. A graphic file depicting a whole page of text can be quite large, perhaps running into the megabytes, depending on resolution and size. But a text file of that same page is unlikely to take up more than a few kilobytes. If the information can be conveyed by pure text, and the font and other formatting are not an integral part of the information, then the document is a good candidate for OCR scanning.

Summary

Scanners have gone from being considered toys to being valuable peripherals for information-related tasks. There are a number of scanner types, and they can perform a number of functions.

Hand-held scanners are compact and inexpensive, but their relatively poor image quality, and the decline in price of better scanner types, has rendered them largely obsolete. Flatbed scanners can be inexpensive, and they have the advantage of being able to scan three-dimensional objects, like books. Their high quality and low price make them an attractive alternative. Sheet-fed scanners yield very high image quality, and these compact scanners take up very little desk space. However, they can only scan single sheets and cannot scan the pages of a book unless the book is disassembled first.

Scanners have a variety of uses, and the most prominent of these is the digitization of images. These images can add impact to Web pages, paper documents, and electronic bibliographic instruction. Scanners can also play an integral role in document preservation, but this must be done with care. Improper scanning can damage fragile documents.

Optical character recognition, or OCR, offers the ability to create editable text files from printed pages that would otherwise have to be retyped to convert them to electronic form. OCR has limitations, but it can still be a powerful tool.

Glossary Terms

OCR
optical character recognition
optical scanner
scan head
scanner
transparent media adapter
TWAIN

9

Power Control Devices

Electricity is the lifeblood of computers, but it can also be a source of great risk. If a machine receives too much or too little power, a variety of problems can result, including lost data or hardware damage. Because of this, a variety of devices have been developed to filter incoming power and ensure that a computer receives the correct amount of clean electricity. Power surges, brownouts, and power failures are the main power problems that can destroy data or damage computers.

Power surges, also called *power spikes,* or just *spikes* or *surges,* are large, sudden increases in the amount of power coming through an electrical line. Large surges can severely damage, or even ruin, a computer. Microchips operate on very small amounts of electrical current and are easily damaged. Because virtually all of the components in a computer are operated by microchips, this leaves all those components vulnerable to surges. Surge damage can range from loss of data to complete destruction of the computer.

Brownouts, also called *undervoltage,* are the other common type of electrical problem. They are opposite phenomena to power surges and occur when the amount of power going to the computer falls below the level needed for the computer to operate properly. While this might seem less of a hazard, it is not necessarily so. Under certain conditions, chips can be damaged by low voltage, and loss of power can cause a computer to reboot, with undesirable consequences. A reboot under those conditions is equivalent to simply turning the computer off without shutting down properly. If users are lucky, only unsaved data will disappear, and everything will operate correctly when they power the computer back up. If they're not lucky, damage can result to the machine, with one of the more common symptoms being corruption of part of the hard drive. While there is some debate about the real effectiveness of some power filters (this will be discussed later), a well-built filter can save a good deal of money and time.

All of these devices work on the same basic principle. They plug into the normal AC wall outlet, and the computer and its peripherals plug into the power filter. Figures 9.1 and 9.2 show a computer without power protection, and one with protection. The filter acts to protect the computer from power surges, and, in some cases, from brownouts or power failure. There are a number of filter types, with varying levels of effectiveness.

One word of advice has to be mentioned before any detailed discussion. *Most of the devices intended to protect a computer from power problems depend on being plugged into a properly grounded outlet.* This usually will be a standard three-pronged outlet. If this basic step is not taken, any other step is likely to be a waste of time and money.

Surge Suppressors

Surge suppressors, or *surge protectors,* are the simplest of the accessories meant to filter the power going to a computer. When they detect a power surge, surge suppressors are supposed to direct it to the electrical ground,[1] pro-

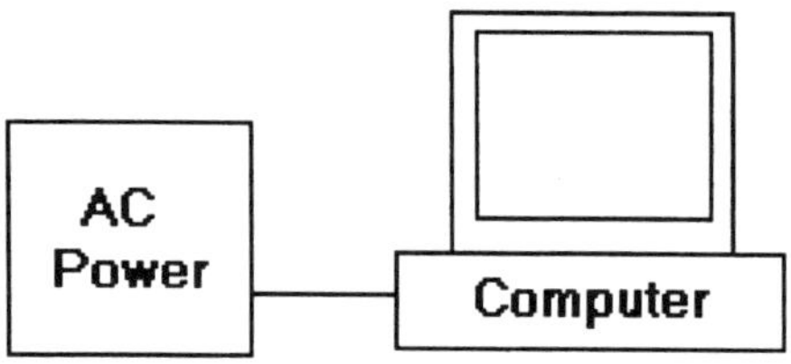

Figure 9.1. A computer with no power filtration.

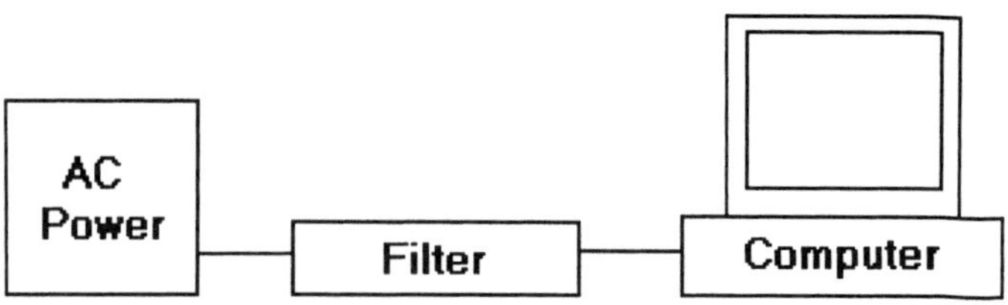

Figure 9.2. A computer with a power filter.

tecting the computer. This only works if the surge suppressor is plugged into a properly grounded outlet. If not, it will not work.

Surge protectors can use a number of components to perform their task, the most common of which is an MOV, or metal-oxide varistor.[2] MOVs have several problems, though. They can handle small surges, but a large surge will not only pass right through an MOV but can destroy its effectiveness, as well. Unfortunately, most surge protectors give no indication that this has happened. If a surge protector's MOVs have been blown, the user, in effect, no longer has a surge suppressor, and the computer is unprotected. For this reason, there is some debate about whether most surge suppressors are worth their cost.[3] They do provide some protection, however, and either a surge suppressor or some other type of power filter should be used for every computer, and for each of the computer peripherals plugged in separately. This also extends to phone lines and network connections. Any electrical wire or cable passing into a computer from the outside is a potential source of power surges.

Surge suppressors are rated in a number of ways. The simplest of these is an indication of how many devices can be plugged into the suppressor, usually at least four. A surge protector should also allow phone cords and network cables to be plugged in.

The more critical measurement of suppressor effectiveness is the voltage level at which the protector will no longer let power pass, the amount of energy the suppressor can absorb, which is measured in joules. The factor to check here is whether the suppressor conforms to the UL 1449 standard dictated by Underwriters Laboratories. A high-quality surge protector will conform to this standard and will have an indicator light to inform the user if the MOVs in the unit have been blown by a large surge.[4]

Power Conditioners

Power conditioners, or *line conditioners,* represent a step up in protection from simple surge suppressors. In addition to protecting a computer from power surges, line conditioners are also capable of bridging small brownouts. Another advantage they have over many surge suppressors is that they can filter out radio frequency interference, or RFI. Radio waves are ubiquitous in our environment and are given off by radios, televisions, and a variety of other devices. Because they are electromagnetic waves, they have the potential to interfere with the operation of a computer. Power cords and phone lines can act as antennae, sending the radio energy into the computer. Some surge suppressors can filter out RFI, but line conditioners are more likely to have that ability.

Backup Power Supplies

Line conditioners and surge protectors share a common fault: in the case of complete power failure, they do nothing to protect the computer or the data. To protect a computer from power failure, a user needs a *backup power supply* that will provide the computer with battery power for a few minutes if the normal power source fails. This is enough time to save data and shut the computer down properly.

There are two basic types of backup power supplies, a *standby power supply,* or *SPS,* and an *uninterruptible power supply,* or *UPS.* They perform the same function, but there are important differences in the way they do it.

An SPS will often perform less reliably than a UPS. This is because an SPS relies on an automatic switch to convert from standard power to battery power in case of a failure. The speed of this switch is critical. If the switch is not fast enough, the effect will be the same as a brownout or short blackout: the computer may shut down improperly and possibly reboot when the power comes back. An outage long enough to make the clock on a VCR blink is more than long enough to cause problems for a computer. A good SPS can switch fast enough to prevent this. The switching time is measured in milliseconds, and 4 ms or less is ideal.[5] The SPS should sound an alarm when power fails.

Uninterruptible power supplies work on a different principle. Unlike an SPS, which provides battery power only when normal power fails, a UPS always runs a computer from its batteries. If normal power fails, the batteries keep operating with no switching time. The UPS accomplishes this by constantly charging the battery from the normal power, and using power from this constantly charged battery to run the computer.[6] A good UPS will also sound an alarm in case of power failure.

Power protection devices are usually much cheaper than the computers they are designed to protect, and they should be considered a necessary accessory. They range in price from hundreds of dollars, for the most elaborate, to less than $50 for the simplest. The choice of which to use depends on the particular computer's function and value to the organization. Inexpensive machines used for public access to the Web, for example, may merit less protection than other machines. If one of these machines is damaged it may be inconvenient, but the overall operation of the library will probably remain unaffected. But if a library operates its own OPAC server, that machine should be considered critical to the very mission of the library. A high-quality SPS or UPS is mandatory for such a machine. Loss of OPAC data can range from being a major inconvenience, if the data is backed up, to a catastrophe if it is not. Larger institutions, with an Information Technology department to maintain the OPAC server, will often relieve the library of this need.

In between nonvital computers and ones that are mission critical, there are bound to be a number of other machines. Part of thoughtful technology planning involves determining how important the machines themselves are to the library and deciding how much value to place on the data they store. For most machines, a good surge suppressor or line conditioner is sufficient, but important hardware and data merit a higher degree of protection.

Other Protective Measures

There are several other simple steps that will enhance a computer's protection from power problems. The simplest has already been mentioned more than once, but it is worth mentioning again. *Use a properly grounded electrical outlet.* If you do not, most filtering devices will be a waste. Many two-pronged outlets are not grounded, and even though three-pronged outlets should be, never assume. Most hardware stores will have inexpensive (probably less than $20) devices for checking an outlet's grounding status. If an outlet is not grounded, contact an electrician about changing it.

Second, do not put a computer on the same circuit as a major appliance, such as a refrigerator or microwave. Notice that when these devices cycle on or off, lights on the same circuit may dim or flicker, and radios or TVs on the same circuit may crackle momentarily. These flickers can damage a computer. While this may not seem particularly relevant to public access computers in libraries, stop and think for a moment about what library staff members might have plugged into outlets in the back office.

Summary

The debate over the value of various types of power protection will, no doubt, continue. But several things are clear.

First, some kind of power filtration is a good idea, because some protection is better than none. A surge suppressor can protect a computer from momentary power overages and may filter out radio frequency interference. A line conditioner will do both as well as bridging small brownouts with steadier power than a simple surge protector.

Backup power supplies provide additional protection by providing emergency power in case of an outage. Standby power supplies do this by switching quickly to battery power after a failure, and uninterruptible power supplies do this by running the computer off batteries that are constantly being recharged.

Glossary Terms

backup power supply
brownout
line conditioner
power conditioner
power spike
power surge
spike
SPS
standby power supply
surge
surge protector
surge suppressor
undervoltage
uninterruptible power supply
UPS

II

OPERATING SYSTEMS AND SOFTWARE

10

Operating Systems

Until now, this book has focused largely on hardware. Indeed, the hardware configuration of a computer tells a great deal about the machine's capacity, its speed, and its eventual obsolescence. But hardware is only one side of the picture. A computer's operating system serves as both a complement to the hardware and as an interface for the user, and it tells a lot about a computer's capabilities.

Definition

An *operating system,* or *OS* (pronounced oh-ess), is the low-level software that takes care of basic system functions such as memory allocation for programs, communication between hardware devices, and allocation of storage space. It also serves as the basic user interface and provides application programs with a standard set of functions for operations such as file storage and hardware control.

If you can describe a computer's hardware configuration and its operating system, you have captured much of what makes a certain model or type of computer what it is. In fact, the term *platform* usually refers to a particular combination of hardware and operating system. IBM-compatibles and Apple computers are two different, well-known platforms. While all IBM-compatibles do not run the same operating system, they all use operating systems from the same family, and the similarities between them far outweigh the differences. The same is true of Apple computers, though almost all of them use some version of the Mac OS, Apple's operating system.

An operating system saves programmers and users a great deal of time and frustration. In fact, computing as we know it would be impossible without operating systems. Without an OS, the authors of each program would have to start from scratch and write a myriad of functions for themselves. Every program would save files differently, use a different method to ask the user for input, and address hardware devices in a different manner.[1] Computers would be difficult to use, at best.

An operating system simplifies these functions, and more, in a variety of ways. For example, when you are saving a file in Microsoft Windows, the dialog box will always look approximately the same, whether you are using a word processor, a Web browser, or a graphics application. This is one of the many ways in which the OS provides a uniform user interface. The user knows what to expect and has a chance to develop a feel for his or her computer's operating system.

An operating system provides hooks, as it were, that attach a program to the system. It provides a relatively uniform look and feel for the user and a uniform development environment for makers of hardware and software. While the installation and configuration of software and hardware may be difficult at times, it would be substantially more difficult without a well-developed operating system.

The uniformity of an operating system does have drawbacks, though. Programs meant for one OS may be totally incompatible with any other system. For example, an IBM-compatible cannot run programs meant for Apples without also running special software to translate the Apple program into a form understandable to the IBM-compatible's operating system. Anyone who has worked in an environment with mixed platforms knows the frustrations this can cause. But however inconvenient these incompatibilities may be, lack of an operating system would make the situation far worse.

While an OS may provide an interface, there are important differences between those two terms. An *interface,* in this context, is a *user interface,* a set of commands, visual aids, or other tools that help the user communicate with the computer. An interface that uses visual elements, such as on-screen buttons to be clicked with a mouse, is referred to as a *graphical user interface* or, more commonly, a *GUI* (pronounced gooey).

Every operating system must provide some interface for the user, but it is not always particularly easy to use. For this reason, an interface is often added as a separate piece of software on top of the OS. The versions of Windows before Windows 95 are an excellent example of an add-on interface. In these earlier versions, MS-DOS, or DOS for short, remained the operating system, and Windows was installed as a separate interface. In its native state, DOS has a *command line interface.* A command line interface receives input from the user as text typed in at a *command prompt.* A command prompt is a set of characters, most commonly C:\> in DOS, followed by a cursor, which indicates that the machine is ready for input from the user. Many users find this cumbersome, and the commands the user must type can be long and difficult to remember. A GUI, like Windows, provides shortcuts so that instead of typing a command, or string of commands, to start a program, the user can simply click an icon. The Mac OS was built as a GUI from the ground up, and almost all functions on a Mac are visually based.

Windows 95 represents a hybrid of both a GUI and a command line interface. It includes the command line interface from DOS but in a different form than earlier DOS versions. Those earlier versions were independent operating systems, but under Windows 95 DOS operates in a *shell.* A shell is a command interpreter that passes commands entered by the user to the operating system.[2] In a certain sense, a GUI, like early versions of Windows, could be considered a shell, but the term shell is usually reserved for command line interfaces. The core of the operating system, to which the commands are passed, is referred to as the *kernel.*

Functions

Operating systems perform a wide variety of functions, many of which remain hidden from the user. But a number of the functions are not hidden and directly affect the user's interaction with the computer.

File Structures

The ability to reliably store and retrieve information is one of a computer's most vital functions. Without it, programs and data files would be no more than ephemera that disappeared when the computer was turned off. An operating system provides a structure in which the user can store data in various formats usable by software.

A file structure in which files were all stored in one large group would be of limited use. The number of files on a computer balloons quickly, posing difficulties for a user trying to pick through them. To avoid this, operating systems allow the user to organize files in *folders* or *directories.* A directory plays the same role as a folder in a file cabinet, serving to organize documents in a manageable order. Only, in this case, instead of paper the documents are made of bits and bytes and can be anything from a simple text file to an executable program file. Directories are often called folders simply because they are so frequently represented by a folder icon in GUIs.

The file system both in IBM-compatibles and in Apples is hierarchical, meaning that directories can contain other directories, which are referred to as *subdirectories.* In most computers with a hierarchical file system, the top-most directory is referred to as the *root directory.* Other than being on top of the hierarchy, the root directory is also distinctive in that it contains certain vital system files. Deleting files from the root directory without being certain what function they perform is never a good idea. The term *root directory* is not generally used in reference to Apples, where the System, Preferences, and Extensions folders together perform roughly the same function.

The file structure of an operating system also gives rise to *naming conventions,* which are protocols for naming files. The most well known of these conventions is the DOS 8×3 naming scheme, which refers to the fact that under DOS, file names can only be eight characters long and can have an *extension* of up to three characters. An extension is part of the file name which tells the OS and the user what type of file it is. For example, in Windows, plain text files have the extension .txt (pronounced *dot tee-ex-tee*), and graphics files stored as bitmap images have the extension .bmp. The dot separates the extension from the main file name. Many users find this restrictive, and so Windows 95 allows longer file names, as the Mac OS always has. In fact, Windows 95 allows file names of up to 255 characters.[3]

Most file naming conventions restrict the use of certain characters. For example, under Windows, a period, or dot, can only appear immediately before the file extension. Placing a dot anywhere else in the file name generates an error message telling the user to rename the file. The operating system's documentation should list the restricted characters, along with any other naming protocols.

Multitasking

Multitasking, the use of more than one program at one time, is one of the most useful functions an operating system can perform. Multitasking was mentioned earlier, but it is worth discussing in more detail. Wise use of multitasking allows users to significantly enhance the performance of their machines, but poor use of this feature can cause system crashes and the frustrations those crashes bring.

Windows 95, the Mac OS, and recent versions of DOS all allow multitasking, with varying levels of reliability and ease. All of these operating systems use *cooperative multitasking.*[4] Cooperative, in this context, refers to the way in which the system attempts to handle the multiple programs. Under cooperative multitasking, each program determines when it needs attention from the CPU and, theoretically, allows other programs to have CPU attention when they need it. But a poorly designed program can take over the CPU, locking out other programs and possibly crashing the computer.[5] Newer operating systems like Windows 95 and recent versions of the Mac OS handle this better than older systems, but they are still subject to occasional problems.

Other systems, such as OS/2 and Unix, use *preemptive multitasking*. Under this form of multitasking, the CPU can decide to stop processing one program and move on to another, regardless of what the first program is requesting. The system is also in charge of making sure that when a program is interrupted, enough information is saved about the state of the program that it can be restarted when the interrupting task is done.[6]

It has to be emphasized that under both forms of multitasking, the switch of the CPU from one program to another is separate from the user choosing which program is the active one. The active program is the one running in the foreground window, and it is the one ready to receive user commands. For example, the user could be browsing on the Web and simultaneously listening to an audio CD. Once the CD is playing, it takes little processing time from the CPU and continues to play in the background. If the program being used to play the CD needs some action performed by the CPU, the switch is performed without the user noticing, in most cases. On the other hand, the user can also make the CD player the active program. If the user wants to skip over a song on the CD, there will be a combination of mouse clicks or key strokes allowing the audio program to become active, while the Web browser continues to work in the background. The user can skip over the song, then switch back to the browser and pick up at the point of interruption. In the meantime, the browser could have continued downloading a Web page uninterrupted. This user-initiated switch is completely separate from any task switching performed by the CPU as it juggles the two programs. At any given moment, either the active or inactive program may be using the CPU.

As mentioned above, operating systems perform multitasking with varying levels of effectiveness. DOS tends to do it poorly, because many DOS programs are written under the assumption that they will be the only one running.[7] Windows 3.1 performs multitasking better than DOS, but it suffers from what many people refer to as a "memory leak." When a program starts, it is allocated a certain amount of memory, and this amount will vary depending on what function the program is performing at any given moment. When the program is done with a particular block of memory, it is supposed to release that block so that it can be reclaimed by the operating system.

Windows 3.1 does not always handle this task properly, and it leaves certain blocks of memory unused even though they are no longer taken up by a program. These ignored blocks gradually grow in number, and the machine runs out of memory just as if it were leaking. Multitasking accelerates this, and Windows 3.1 will frequently slow down and then crash if the user is switching frequently between programs. If a noticeable slowing of the computer is detected while multitasking under Windows 3.1, users should save their work, exit all programs, and reboot the machine. This will restore all the memory.

Windows 95 and the Mac OS do a much better job of memory allocation, and users will experience far fewer problems with multitasking under these systems. However, multitasking effectively on any system depends on the amount of RAM available. Multitasking can be performed on 4 MB or less, but with difficulty; 16–32 MB are more realistic amounts for multitasking with recent applications. If a program takes up a large enough amount of RAM, and sophisticated graphics programs often do this, multitasking may not be practical. Some memory-hungry programs simply run better by themselves.

Hardware and Software Interrupts

Computers are not so much discrete machines as much as they are assemblages of various components, both hardware and software. These components make use of the same memory and processing resources, and the operating system must allocate these resources in a way that allows the whole system to function smoothly.

In the chapter on expansion cards, interrupts were discussed in regard to hardware. Just as a device like an expansion card can send an interrupt to the CPU, software can as well. The most important task the OS performs in this context is to make sure that any process, whether it involves hardware or software, is interrupted in an orderly way so that it can resume after the task causing the interrupt has finished. Without this capability, a machine would only be able to function in a linear progression without any variation.

A Taxonomy of Operating Systems

Operating systems have evolved greatly in the last several decades. The popularization of computers has created a drive to make the systems less arcane than in the past, and the result has been progressively more usable machines. But some of the features of the older systems remain embedded in current operating systems, and this is especially true in IBM-compatibles. Knowing how operating systems evolved can help explain some of the quirks and shortcomings that the user will inevitably encounter.

MS-DOS

MS-DOS, short for Microsoft Disk Operating System, is the original operating system for IBM PCs and compatible machines. It was originally referred to as a disk operating system because many of its early functions dealt with storage of data on disks. Called DOS for short, it still underlies all versions of Windows.

MS-DOS was developed in 1981 as a 16-bit operating system for IBM personal computers. DOS quickly overwhelmed its competitors, establishing itself as the standard for IBM-compatibles.[8] Some IBM machines have been sold with PC-DOS, a version of DOS that is functionally the same as MS-DOS, with only minor differences.

It is interesting to note that a number of operating systems have used the term DOS, including AmigaDOS and the original Apple DOS. But none of these systems ever gained the prominence of MS-DOS. For this reason, when someone refers to a DOS machine or *DOS computer,* they are almost always referring to an IBM-compatible running some version of Microsoft's MS-DOS.[9]

Recent versions of DOS offer some multitasking capabilities, but it remains, for the most part, an operating system meant to run one program at a time. The biggest exception to this is the use of *TSRs,* or *Terminate and Stay Resident* programs. TSRs are utilities that function periodically but remain in memory when they are not running. This is in contrast to most programs that are emptied from memory when not in use. Screensavers and antiviral programs are the best examples of TSRs. A screensaver, for example, remains in memory, waiting for one of two events. Either the computer remains idle for a given period of time, after which the screensaver starts, or the user enters a certain combination of characters from the keyboard to start the screensaver manually. Screensavers are much more common in graphical environments, like Windows or the Mac OS, than in DOS, but the principal behind any TSR is the same.

TSRs are useful, but they can also cause conflicts and take up valuable memory. For this reason, they should be used judiciously. Because of the conflicts they can cause, TSRs should always be suspect when a machine is experiencing memory problems or unexplained crashes. Turning off TSRs to see if the problem stops is a good first step in troubleshooting these sorts of problems.

Memory management is one of the most frequently encountered problems in DOS. DOS is structured so that no matter how much RAM a machine has, programs run in the first 640k, referred to as *conventional memory.* TSRs and various utilities eat into this memory, and sometimes this leaves insufficient memory for applications to run. There are memory management utilities that alter the way DOS uses memory, and these utilities are sometimes necessary to solve these conflicts. Because Windows through version 3.11 is an interface that sits on top of DOS, those versions of Windows inherit this kind of problem. Fortunately, much of the problem is eliminated under Windows 95.

DOS is made up of a large number of files that together constitute the operating system, but two of them bear separate explanation. Autoexec.bat and Config.sys work in tandem and determine much of a system's configuration. Both files reside in the root directory and *should never be moved to another directory or deleted.* They are among the first files read or executed by the computer during startup, and absence of either one can effectively shut the computer down.

Config.sys is a text file containing information about the computer's configuration,[10] and it contains information about both hardware and software. It runs before Autoexec.bat. If Config.sys is deleted or corrupted, devices may not run properly.

Autoexec.bat is run after Config.sys is read, and it contains a number of commands for performing basic functions, such as turning on antiviral TSRs and allocating system resources.

Autoexec.bat is a very specialized *batch file,* indicated by the .bat extension. A DOS batch file is a file containing a list, or batch, of DOS commands. Two such DOS commands, Defrag and Scandisk, were discussed in the section on maintenance of hard disk drives. A user could, rather than remembering the individual commands, combine them in a batch file. When the batch file is run, the commands would be executed one after another without the user having to enter them separately.[11]

Autoexec.bat uses a long string of commands to perform system functions. This file can be edited, but great care needs to be taken. Often, when a DOS program is being installed, the user will be prompted to approve changes the program is about to make to Autoexec.bat or Config.sys, or both. Programs generally do this properly, and letting the program make the changes is easier than making them manually. Most programs will make a backup copy of the file if they change it. Make careful note of this backup copy's name. If there are system problems after the change, the old configuration can be restored by changing the name of the backup file back to Autoexec.bat or Config.sys, whichever is appropriate. The new program may not run after this restoration, but the rest of the system should run as well as it did before. Troubleshooting these conflicts can be extremely difficult, and most users will probably want help in editing the newer version to run properly.

Windows

All of the versions of Windows through 3.11 had the same basic structure. As mentioned before, they were designed as an interface to sit on top of DOS, creating convenient shortcuts and a more comfortable environment for users. Because of this relationship they bore to DOS, they inherited its virtues and shortcomings. As a result, much of the discussion about DOS also applies to these early versions of Windows. In this context, early will refer to all versions of Windows before Windows 95, and Windows NT will be discussed separately.

Early versions of Windows did improve on DOS in a number of ways. One of the most significant of these was improved multitasking. Beyond TSRs, Windows allows the user to have more than one application open and running, and switching between these multiple applications is quite easy and involves only a few keystrokes or mouse clicks. This vastly simplifies and speeds tasks like editing a Web page. The user can have an editor open for the HTML text, and can run a Web browser at the same time to see work as it would actually appear on the Web. While an increasing number of Web editors allows previews without the use of a browser, not all do, and this sort of multitasking can make computing much more pleasant. Unfortunately, the infamous memory leak in early Windows versions detracts somewhat from this feature.

The other major advantage of Windows is the use of visual icons instead of typed commands to perform system functions. Instead of having to type in sometimes cumbersome commands to switch to a certain directory, then often another command to start an application, this can be accomplished by clicking an icon on the screen. The Program Manager is a program within Windows that displays these icons in separate program groups, which are groups of related applications. Figure 10.1 shows the Windows Program Manager with one of its program groups open. To many users, this is no doubt a familiar view.

Some additional system files are used in Windows, but the underlying structure is essentially the same as DOS. Autoexec.bat and Config.sys are still vital to system performance, and many maintenance features, such as Defrag and Scandisk, are still run through DOS. There is one wrinkle that

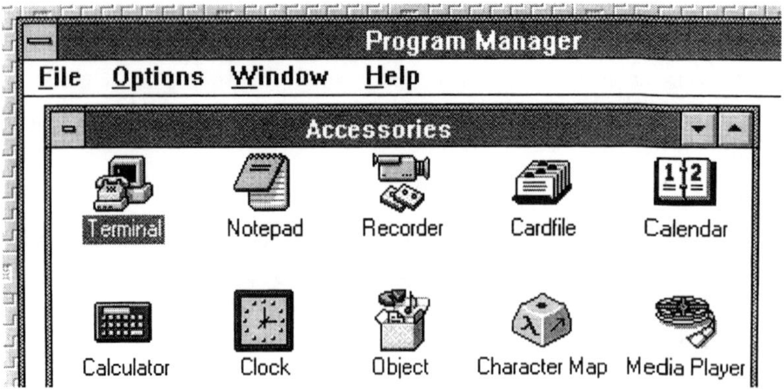

Figure 10.1. Windows 3.1 Program Manager.

users need to keep in mind, though. There are two ways to use DOS under early versions of Windows, and they are *not* equivalent. In one of the program groups within the Program Manager, there is an icon for DOS. If users click on this icon, they are taken into a DOS *shell*, not into pure DOS. For many functions this makes no difference. But for others, it does. Some memory-hungry DOS programs will not run properly, or at all, in a shell. To run them, the users must exit Windows completely so that they are in DOS. The same is also true for Defrag.[12] In general, it is a good idea to run DOS programs while actually in DOS, rather than from the shell.

Windows 95, or Win95 for short, is a more thorough makeover of the Windows system than any of the previous versions. While there is some dispute over whether Windows 95 is a unified operating system or still an interface running on top of DOS, Microsoft considers it a unified system, and it behaves as if it is. To give an example, if the user goes into the Windows 95 DOS shell and types the DEFRAG command, the machine will come back out of DOS and run a newer, Windows version of the utility.

The file structure underlying Win95 is essentially the same as DOS, but with some enhancements. One of the more prominent enhancements is the ability to create long file names, which, for the most part, does away with the 8×3 naming restriction. Technically, a file given a long name in Win95 is given two names: the long name, which displays fully in Windows, and a shorter name that appears in the DOS shell and conforms to the older 8×3 protocol. This quirk can cause confusion for a user unaccustomed to it. If, for example, a user names a file "ReallyBigFile.txt" in Windows, that name will display in dialog boxes and in file management utilities. But if the user goes into a DOS shell and looks for the same file, it will have a truncated name, something like "Really~1.txt." The user must refer to this truncated name when in a DOS shell.

Windows 95 has another quirk worth noting. It is predominantly a 32-bit operating system but still has some 16-bit elements. The 16-bit code remains so that Win95 can maintain backward compatibility with programs written for Windows 3.1. Backward compatibility is the ability to use software or hardware designed for an earlier system. Without this sort of compatibility, every time an operating system was upgraded, users would have to discard their old software and buy new versions. But because Win95 is backward compatible with almost all Windows 3.1 applications, users can bring those applications with them when they upgrade.

There is a downside to backward compatibility, though. While Win95 manages memory and other system resources better than Windows 3.1, and is much more stable, the retention of 16-bit code also means that some of the problems remain. If Win95 were fully 32-bit, and running only 32-bit applications, then it would be even more stable. But because of the 16-bit code necessary for backward compatibility, Win95 is prone to some of the same crashes that occur in earlier versions of Windows. At some point in the future, compatibility with 16-bit applications will fade in importance, and Windows will probably become a fully 32-bit operating system.[13]

Windows 95 also features a substantially different interface than Windows 3.1. Rather than the old Program Manager with its program groups, Win95 has a Start button that brings up a list of program groups and individual programs as well as a "Desktop" on which icons for either applications or individual files can be placed. Figure 10.2 is a partial screenshot from Windows 95, showing the program groups listed under the Start button, and with program icons also visible on the Desktop in the background. Other than an easier interface, Win95 also has greater speed than Windows 3.1 could achieve.

There are also structural differences between Windows 95 and earlier versions. One of the main differences is Win95's use of a *registry.* The registry is a database of system information similar to the information in the old Autoexec.bat and Config.sys files. It is more difficult to modify or edit, though, and most users will probably want to avoid that task. Autoexec.bat and Config.sys still exist under Win95, but their role is vastly reduced. Some programs still make use of them, but most do not.

Windows 98, an upgrade to Windows 95, is being released as this book goes to press. The preliminary release version has received mixed reviews,[14] and there are a number of things to consider when deciding whether to upgrade. The primary consideration arises from the fact that Win98 will be the last

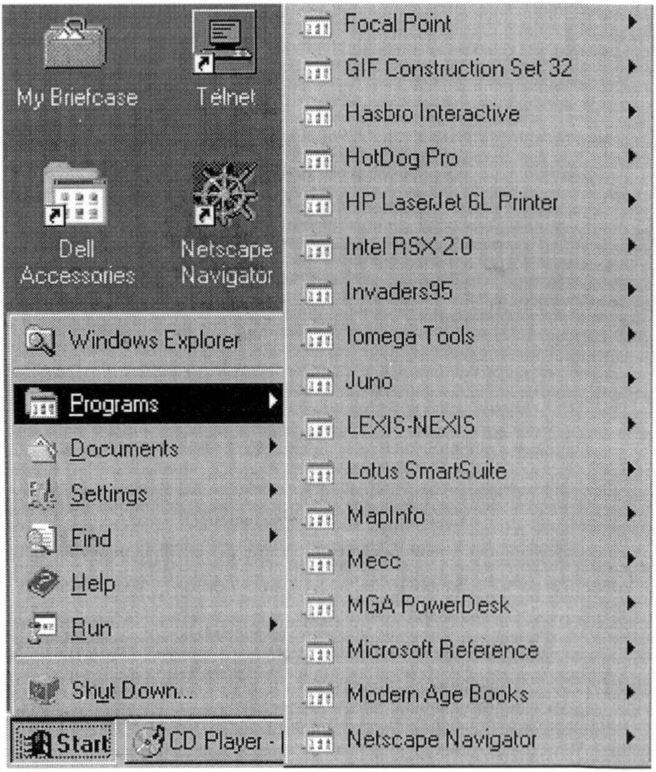

Figure 10.2. Windows 95 Program Groups and Icons.

upgrade to the Win9x OS family.[15] After Win98, Win9x and Windows NT (see below) will merge into a single operating system. The extent to which that combined operating system will be compatible with older applications and hardware is uncertain. In addition, Win98 does not make extensive changes to the operating system. Its principal benefits are the fixing of bugs from Win95, better integration of the Internet and the local desktop, and minor performance enhancements. It will be a useful upgrade, but not a necessary one, especially given the approaching end of the Windows 9x OS family.

Windows NT is a step up even from Windows 95 and has had the most complicated evolution of all the Windows systems. It actually developed from a version of the OS/2 operating system. OS/2 was initially a cooperative effort between Microsoft and IBM, but the companies halted joint development and developed the systems separately.[16]

NT, as it is commonly known, is a fully 32-bit operating system. The most recent version, 4.0, has the same interface as Windows 95, and earlier versions had an interface similar to Windows 3.1. The similarities largely end there, though. Because NT is a 32-bit system and has excellent resource management, it is extremely stable. These 32-bit applications run in their own block of protected memory, which cannot be used by any other program. Conversely, if a program running in protected memory crashes, it is much less likely to lock up the whole system than a program crashing under Windows 3.1 or 95.

This stability comes at a price, though. Windows NT is not backward compatible with other Windows versions, and it will only run 32-bit programs.[17] Users who want to switch to NT must discard much of their old software and start over. Some Windows 95 programs will run under NT, but many will not. For this reason, the decision to switch to NT is a significant one.

Eventually there may be a single version of Windows, but it is not here yet. NT is meant for high-end computers often referred to as *workstations,* for *servers,* and for machines used on some large networks. *Workstation* is used to refer to machines that have more power than the average personal computer and are used for applications, such as complex graphics, that would severely strain a lesser machine. A server is simply a machine whose primary function is to hold information meant to be accessed over a network. There are, in fact, two versions of NT. The first, referred to as NT Workstation, is meant for individual workstations. The other, NT Server, is meant to operate on machines used as servers. NT Server is significantly more complex, and expensive, than NT Workstation and is more than most users need.

Because use of NT brings along with it the complications of software incompatibility, and because it is a more complex operating system to use than Win95, many libraries do not use it unless they are part of a network that requires or suggests it. Win95, for most users, makes just as good an OS on a network.

There is also a third, less widely used version of Windows, Windows CE. CE is designed for *palmtop* computers and *embedded systems.* Palmtops are small computers that have limited hardware capabilities and run applications that are generally scaled down versions of full applications such as word processors or spreadsheets. Palmtops should not be confused with personal digital assistants, or PDAs. PDAs perform very simple functions like organizing names and addresses, and generally have little or no ability to handle complex data. An embedded system is one meant to run in a very specific environment, such as in a stand-alone kiosk, an automatic teller machine, or industrial machinery.

Palmtops and similar devices are becoming more common, and this may give Windows CE an additional boost in prominence with the impending merge of Windows 95 and Windows NT. Instead of offering three operating systems, Windows CE, Win95/98, and NT, Microsoft may offer only two, Windows CE and the combined operating system resulting from the merging of Win95/98 and NT.

Upgrades

Like any other piece of software, operating systems are upgraded periodically. But unlike most software, the decision to upgrade an operating system is not always simple. Existing software may not be completely compatible with the new OS, and all the users on a library staff, and the patrons, as well, will have to learn new procedures. This can be complicated enough with a single application, but changing the entire OS creates even more complications. Nevertheless, OS upgrades are inevitable. As applications are upgraded, they often require newer versions of an operating system, and the features of new OS versions become de facto standards.

Upgrading versions of DOS, and versions of Windows before Win95, is relatively simple, although as DOS and early versions of Windows fade from prominence this will be a less common occurrence. The upgrade will generally install from CD or diskettes like any other software. There are two general rules to keep in mind. First, while an OS upgrade should leave data undisturbed, it is wise to backup important data before performing the upgrade. If anything goes wrong, there is some potential for data loss. This potential is generally small but exists nonetheless. The other rule of thumb is that all TSRs, especially antiviral programs, should be turned off before the upgrade begins. This applies to screen savers and any add-on shells that are used on the machine. These two precautions apply to any OS upgrade.

Unfortunately, turning off TSRs is not always easy. In the case of a screen saver, it may be as simple as opening up the program and clicking an icon to turn it off, then rebooting the machine. But some TSRs are launched by Autoexec.bat or Config.sys, and the user must edit those files to turn the TSRs off. It is imperative when editing these files to save the old version under a different name in case of trouble. If something goes wrong, you may be very glad to have the old version to restore the machine's former settings.

Editing Autoexec.bat or Config.sys to turn off TSRs is easy in principle, but difficult in practice. Both files execute line by line during startup, and turning off a TSR is as simple as stopping the execution of the line that calls for the TSR to start. This is done by adding the term *rem* to the beginning of the line, saving the file, then rebooting the computer when all the changes have been made. REM (pronounced *rem*) is short for remark, and designates the line as a remark rather than a command. Remarks are put into batch and system files to provide explanatory information to a person looking at the file, and they are ignored by the computer. Doing this is often referred to as "remming out a line." This sounds simple in theory, but the difficulty lies in deciding which line to rem out. Under DOS and Windows 3.1, Autoexec.bat and Config.sys can contain dozens of lines whose purpose is uncertain to a user not used to looking at them. If a user feels uncertain about these steps, it would be best to find a good DOS reference manual or a colleague who can help him or her.

Upgrading from Windows 3.1 to Windows 95 is only slightly more complicated. In this case, the advice to shut off any TSRs is especially important. Windows 95 makes some fairly drastic changes when it is installed, and the operations performed by many TSRs can interfere. Once the upgrade is started from the CD-ROM, it usually runs smoothly. The upgrade program saves all of the old system settings and information, and this may become important. If the upgrade does not go well, the CD contains an uninstall utility that will remove all of the Windows 95 elements that were installed and restore all of the former system settings.

There is no upgrade available to switch to Windows NT from Windows 95 or earlier versions. Users of these systems must start from scratch as if the machine were new and having an OS installed for the first time. The only upgrades available to Windows NT are from earlier NT version.

The Mac OS

The Mac OS is Apple's proprietary operating system, designed to run on its own computers. It popularized the use of GUIs, and the market pressure this created has often been said to have forced Microsoft's move to the Windows environment. The Mac OS is a 32-bit operating system built from the ground up as a GUI. It has no command line interface to speak of and operates completely by means of icons and dialog boxes. The current version is Mac OS 8 and an upgrade named Rhapsody is forthcoming. At least through OS 8, the system maintains extensive backward compatibility with older Mac applications.

The Mac OS uses a storage system of folders, and folders within folders, just like its Windows and DOS counterparts. The interface also centers around a virtual desktop.

There are important differences in system design, though, between Apples and IBM-compatibles. There is no equivalent to Autoexec.bat and Config.sys, and any equivalent to the Win95 registry is largely hidden from the user. The heart of the system resides in the System folder, and most configuration information is kept either there or in the Preferences or Extensions folders. Software installation is often as simple as copying configuration information to either or both of these folders and copying a separate program folder to the hard drive.

Extensions, also called *inits,* are an important part of the Mac OS. They play a number of roles, but, in general, they are pieces of software meant to extend the computer's capabilities in some way. For example, they may play the role of a device driver to enable the computer to use a CD-ROM drive, or they may add a capability, like voice synthesis. Installation of extensions is generally straightforward, and usually involves copying the extension from diskette or CD to the Extensions folder. Unfortunately, extensions can conflict with each other, and the conflicts can be difficult to diagnose. These conflicts can slow the system down, or cause periodic system crashes or seemingly random errors. If a library will be using Apples, it should also purchase software designed to diagnose such conflicts, such as Casady and Greene's *Conflict Catcher.*

The Apple file system, while superficially similar to Windows, has structural differences. The Mac OS stores files in two *forks,* or sections. One fork, the *data fork,* contains the data that any system would need from the file. The other fork, the *resource fork,* stores information about the file's icon, and other information about how the file will interact with the system. When a Mac file is transferred to another system, the resource fork is usually left behind, and only the data fork is transferred.[18] The exception to this comes when an Apple reads an IBM-compatible disk. If the disk is used in an Apple, then used in an IBM-compatible, the user will find an empty file called "resource.frk," or some abbreviated version of that name, on the disk. This is the remnant of the resource fork created by the Apple.

It is important to note that even though Apples read IBM-compatible disks, it does not necessarily mean that they can work with the data on the disk. If the data is in a standard format, such as a .gif file for graphics or a .txt file for text, then the Apple will be able to read the file with no trouble. But it will not be able to execute any program files on the disk, or read files in special formats, without additional software allowing it to do so.

Nonstandard Operating Systems

Defining nonstandard with respect to operating systems can be difficult and is truly a matter of perspective. But this book focuses on desktop computers used in libraries, the vast majority of which use either some version of Windows or the Mac OS. For this reason, nonstandard will be defined, for our purposes, as any other operating system. Be aware, though, that in a different context, Windows and the Mac OS would be considered the nonstandard systems. For example, in certain networked environments, or environments where certain specialty applications predominate, Unix would be considered the standard. These other operating systems do have some relevance to libraries, though, and they deserve mention.

UNIX

UNIX is as much a cultural phenomenon as an operating system. Experienced UNIX users are often proud of their comfort with the system and often consider UNIX to be vastly superior to other systems. Without doubt, UNIX is significantly more complex, and has more capabilities, than Windows or the Mac OS. UNIX has built-in programming languages, such as C, and has a variety of shells for the user. It is also an excellent Operating System for networking, since it has a large number of utilities for that purpose.

In library settings, UNIX is most often encountered as the Operating System on an OPAC, Web, or e-mail server. There are GUIs available for UNIX, with one called simply X being the most common. Usually, though, UNIX is used from a command line interface. The commands are superficially similar to DOS commands, but there are many more commands under UNIX, and they tend to be much more flexible, with a large number of options available for each. It also has a hierarchical file structure like DOS, with directories and subdirectories.

Unfortunately, an explanation of even the basic commands in UNIX is beyond the scope of this book. Those who use a UNIX system for work can often get help from a colleague, and there are a number of excellent books available to introduce the operating system.

OS/2

OS/2 is a proprietary operating system sold by IBM. As mentioned above, Windows NT was originally an offshoot of a cooperative effort by IBM and Microsoft to develop OS/2, but any relationship between the two systems is a distant one at this point.[19] Despite the fact that it will run Windows programs, OS/2 has never caught on, and is rarely encountered. It is worth mentioning it, though, because some IBM computers have shipped with OS/2 installed, though they generally come with some version of Windows.

Summary

Operating systems are the low-level software that perform many system functions. These systems serve as a complement to hardware, and the OS and hardware together largely define a computer's capabilities.

An operating system will offer an interface for the user. This may be a command interface, under which the user must type in commands, or a graphical interface, which allows the user to click on-screen icons instead of typing commands. The interface may be separate from the OS, as in early versions of Windows running on top of DOS, or it may be an integrated part of the operating system, as with Windows 95 and NT, and the Mac OS on Apples.

Multitasking operating systems allow the user to run more than one application at a time, but they do so with varying levels of effectiveness. Some applications require a large amount of memory if they are run in a multitasking environment.

Upgrading operating systems is not extremely difficult, but certain rules of thumb must be followed. The most important of these is that all TSRs must be turned off before the upgrade proceeds.

DOS, Windows 3.1 and 95, and the Mac OS account for the majority of operating systems used by librarians. Other systems, including Windows NT, UNIX, and OS/2, are seen occasionally but are much less common.

Glossary Terms

batch file
command line interface
command prompt
conventional memory
cooperative multitasking
data fork
directory
DOS computer
embedded system
extension
folder
fork
graphical user interface
GUI
init
interface
kernel
multitasking
naming convention
operating system
OS
palmtop
platform
preemptive multitasking
registry
resource fork
root directory
server
shell
subdirectory
terminate and stay resident
TSR
user interface
workstation

11

Software

While hardware and an operating system together constitute a stable, usable platform for computing, without additional software a computer can do very little. The operating system's protocols for storing files, and ability to manage system functions, and the hardware's ability to produce a sharp image on the monitor or high-quality printouts would all be of little use without software tailored for specific tasks.

Applications

When we talk about software, more often than not we are talking about *applications.* An application, or *app,* is a program designed for a specific function or range of functions. Word processors, Web browsers, spreadsheets, HTML editors are all examples of application programs. Applications turn a computer from a simple computing device into a useful machine.

People often use the term application in more than one sense, though. Beyond referring to programs meant for specific tasks, computers have more general applications. For instance, creation of Web pages is an application of computers in general, and in that sort of context, application has a meaning more akin to use. In this book, *use* will designate a general type of work on a computer, and *application* will denote specific programs.

Productivity Applications

Applications come in a variety of types, but a few are central to library use. The most ubiquitous are *productivity applications,* sometimes referred to as *office applications.* These applications perform the bread-and-butter functions of many computers: word processing, spreadsheets, databases, scheduling, and so on. They are the most common applications, and an increasing number of computer makers include them on their machines as a matter or course. They are a necessity for library staff and, in many academic libraries with computer labs, for patrons as well. Productivity applications often come as a *bundle* or *suite,* which is a group of related applications that are sold together. They often work well with each other, and data can be exchanged between them more easily than between applications from different companies. Microsoft Office is perhaps the best known example of a productivity suite, though there are similar packages available from Lotus, Corel, and other companies. Use of a suite makes it very easy to, for example, build a spreadsheet, then include that spreadsheet in a word-processed document.

Many low price computers do not include productivity applications. Buying the applications bundled with the computer is almost always cheaper than adding them later, but many public-access PCs have no need for office applications. If the machine is used for Web and OPAC access, it would be a decided disadvantage to have patrons doing spreadsheets or word processing while other patrons waited for access. In that case, a computer that included productivity applications would be a liability.

Cataloging Applications

After common productivity applications, various types of cataloging applications are probably of most use to librarians. OPAC software has become the heart of many libraries, as OPACs offer not only access to local holdings but to remote holdings as well.

OPAC software comes in a variety of types, from versions meant to run on single computers for small libraries, to large, complex versions meant to run on servers accessed over a campus or other large network. Most OPACs are networked, and work on the *client/server model.* Under this model, a server acts as the central repository for all the cataloging information, and individual machines on the network act as *clients,* that is machines that access the information and display it for the user. It is important to note that the client/server model applies to many, many applications other than OPACs. It is the basic model behind the Internet, among other things.

For most library purposes, client computers can be divided into two groups: *dumb terminals* and PCs running client software as an application.

Dumb terminals consist of a display screen and keyboard and do no processing of their own. They communicate with the server, which does the actual processing, and the dumb terminal simply displays the results. They are generally capable of text only, with no graphics. These terminals are being superseded by PCs, which not only run client software for the OPAC, but also perform a number of other functions, such as Web browsing. Many OPAC servers now provide information over the Web in HTML format, and a Web browser, like Netscape or Internet Explorer, can serve as the client software. This is an excellent way to integrate OPAC information with information from outside electronic sources.

It is also important to note that HTML-formatted information does not have to be delivered over the Internet itself. This is also a very effective way to deliver information over local networks, or even on disk. This will be discussed more thoroughly in the section on networks.

Server software comes in a variety of configurations. Some server applications are meant to run on large, UNIX-based servers. These versions may require the support of an Information Technology department staffed by nonlibrarians. These systems are sufficiently complex so that administering them ceases to be library work and should be considered more appropriate for a pure computer professional. Smaller server applications that run on a server using an OS like Windows NT may be well within the capability of some library staff members. This is another planning issue that must be deliberated by each individual institution. The skills of a librarian and the skills of a computer specialist run together when it comes to maintenance of an OPAC server. In many libraries, especially large academic ones, server administration is done by a specialist outside the library, and the librarians who perform technology functions within the library, such as cataloging or instruction in the use of electronic resources, act as liaisons with the specialist.

OPAC software is often separate from a variety of applications used in actual cataloging. If a library connects to a bibliographic utility, like OCLC, then separate software will be needed. These applications will include a combination of network functions to connect to the bibliographic utility and a variety of capabilities for manipulating and editing MARC records. This type of software is meant to run on a PC or workstation, and there are generally a number of methods for connecting to a bibliographic utility over the Internet, by dialing in directly, or by connecting via a local network.

An increasing amount of cataloging software is appearing on CD-ROM, including items that librarians normally associate with cumbersome print volumes, like the *Library of Congress Subject Headings.* The availability of these items electronically has given rise to the idea of the *cataloging workstation.* These workstations provide simultaneous access to multiple CD-ROMs as well as networked resources like bibliographic databases. The idea is that catalogers can find all the resources they need without leaving their computers, saving time and streamlining their workload. The ability to access multiple CDs from a single workstation will be discussed in the chapter on networking.

Network-Related Applications

Libraries deal with an increasing amount of networked information, both through local networks and wide area networks, like the Internet. The software and hardware needed to build or access a network will be discussed later, but there are network-related applications that should be discussed along with other types of software.

Many libraries are mounting pages on the World Wide Web, and any institution doing this should make use of an HTML editor. *HTML,* or *Hypertext Markup Language,* is the language used to design Web pages. A plain text editor can be used to produce Web pages with HTML, but an editor built specifically for that purpose will be more effective. It will offer shortcuts for inserting HTML commands into documents, commands that enable users to produce effective pages. Some of these editors require a separate browser to view the work, but an increasing number of HTML editors offer *WYSIWYG* capabilities. WYSIWYG is an acronym for "What you see is what you get," and denotes editors that display the page being created as it would appear on the Internet. Non-WYSIWYG editors display the HTML codes, which are normally invisible to the user over the Internet since they are used by the computer to lay out the page and are not displayed on the screen. To put WYSIWYG in a different perspective, it is a capability that almost any word processor will now possess. If text is put in italics like *this,* we only see the italicized text on the screen, not the command telling the computer to display it that way. In a non-WYSIWYG editor, we would see something like <em>this</em>, and we would need separate browser software to display the actual italicized text. WYSIWYG editors are often criticized, though, for producing bloated, inefficient code that results in slower downloads than code written by a person using a simpler editor. Machine-produced HTML can also be more difficult to edit by hand later. The obvious advantage to a WYSIWYG editor is that staff members do not need to learn HTML, which is a worthwhile consideration. Many recent word processors also have the ability to save files in HTML format, and they do a creditable job. With the increasing prevalence of Web pages for libraries, though, some piece of software capable of producing HTML is becoming a necessity.

Graphics Applications

Graphics applications represent another technology that has gone from being a luxury to being a mainstream, nearly necessary application. Information is being represented visually with increasing frequency as software makes it easier to do so. Web pages, graphic presentations, and paper handouts are just a few examples of media that can incorporate graphic elements. Not every library deals with graphics extensively enough to need an advanced image-editing program, but at least a simple graphics editor should be included in a library's selection.

Graphics applications can perform a number of operations on pictures. The size of the picture can be altered, as well as brightness and contrast and other parameters. An optical scanner and image software together enable a user to scan photographs for incorporation into Web pages or presentations and allow the editing of smaller pictures for use as icons.

Many computers now come with simple graphics programs already loaded, but these may not be adequate. Graphics files can be saved in a variety of file formats, and the simple image editors included with many computers will only work with one or two formats. At a minimum, a graphics application needs to be able to work with three graphics formats: Graphics Interchange Format (.gif files, pronounced *gif,* occasionally *jif*), Joint Photographic Experts Group (.jpg or .jpeg files, pronounced *jay peg*), and bitmap files (.bmp). There are many formats, but these three represent the basics. Graphics on the Web are usually in either GIF or JPEG format.

Statistical Applications

Statistical applications are also growing in importance for libraries. Collection assessments and user surveys are examples of cases where statistics can be a valuable tool. Fortunately, many libraries already possess basic statistical software, perhaps without knowing it. Most productivity suites come with a spreadsheet application, and spreadsheets often have statistical features adequate for library needs. These spreadsheet functions should be explored before specific statistical software is considered. Statistical applications are not generally taxing on the computer, but they can be expensive and difficult to learn. Because spreadsheets have wide applications for budgets and planning, they can be a wiser investment of staff time than a full-fledged statistical package with powerful features but narrow application.

Programming Applications

An increasing number of Web pages sports features created using programming languages like Perl or Java. These languages can be used for anything from animating icons to creating complete programs, such as games or interactive forms to be filled out and submitted over the Internet. The two languages are fundamentally different, and different resources are needed to use them.

Perl is a part of UNIX and, strictly speaking, needs no separate application for use. Perl is usually programmed from within a UNIX shell, and access to Perl would be a matter to take up with the administrator of a UNIX server.

Java, on the other hand, does require a separate application. Generally this will come in the form of an *IDE,* or *integrated development environment.* This is sometimes referred to as an *interactive development environment.* An IDE is actually a closely integrated system of applications for creating programs in a given language. It will include editing features and capabilities for diagnosing errors in a program.[1] A Java IDE can range widely in price up to a few hundred dollars.

Before investing even in a modest programming application, though, a user needs to be aware that learning even the basics of a programming language can be time consuming. In the case of learning a programming language used to enhance Web pages, it is very much open to question whether it is worth the effort for most libraries. Most of the information presented on library pages is basic textual or graphic material for which enhancements can actually be a distraction. Learning a programming language is well within the capabilities of most librarians, but the investment of time may not be a wise one.

Selection and Installation of Applications

Selection of software can be a surprisingly complicated task. It is also a task where knowledge of hardware is a distinct advantage. Almost any piece of software will come with a list of *system requirements,* which is a list of minimal hardware and operating system capabilities a computer must have to run the program. It should be stressed that the requirements listed on many software packages are *minimum* requirements. A machine that just barely meets these requirements may run the software but might do so slowly or poorly. Recently, some software makers have begun to list minimum and a set of higher, suggested requirements. The suggested requirements should be taken as the more realistic list.

The following is a list of common elements in system requirement lists and an example of what each requirement might be for a hypothetical program.

- CPU: Pentium or greater
- Operating system: Windows 95 or NT
- RAM: 16 MB or better
- CD-ROM drive: 2x or better
- Hard drive space: 50 MB or more free
- Sound card: Sound Blaster compatible

A user should never assume a machine is capable of running a particular program. Applications can have surprising requirements, and if a machine for which the program was bought does not meet the requirements, it can be an expensive mistake.

Licenses

Planning for software purchase must involve discussion of *licenses.* A license is a legal agreement that comes with the software and dictates such limits as the number of machines on which the user can install the software, or how many users may use the software at any one time. Because of potential copyright infringements, licenses are becoming a significant issue for libraries.

Software sold on the retail level usually comes with a *single-user license,* meaning the software can be used by one person, and this usually applies to software included with a computer purchase. A user cannot legally copy it to other machines. Many institutions respond to these issues by buying *multi-user licenses* or *site licenses.* A multi-user license allows installation of the software on multiple machines, and the license will indicate either the number of machines on which the application can be installed, or the maximum number of simultaneous users. Multi-user licenses may be sold either in standard increments, such as five users, ten users, and so on, or there may be a flat fee for the software with a single-user license, plus a fee for each additional user.

A site license is a flat fee, and allows any user on site to load and use the software. Site licenses are common in academic and other large institutions. Licensing agreements often allow the institution to buy inexpensive but high-quality computers that come without software. The software is then installed on site, at a price even lower than the bundled price. Any library or information center with a large number of computers should weigh different licensing schemes when budgeting for technology.

Application Installation

Program installation seems to become easier every year as operating systems and applications evolve. There are still potential problems, though, and a few rules of thumb should be remembered.

One rule applies to almost every area of computing, but it is especially relevant here. *Read the documentation.* Users can avoid many unpleasant surprises by taking this step. Most installation procedures are quite straightforward, but there can be unusual steps required by a particular piece of software. A botched installation can be time-consuming and expensive in staff time.

The other rule of thumb is to shut down all other programs before installing a new one. Installing a piece of software while others are running can create conflicts, especially if the running program and the one being installed use the same system resources. Generally, TSRs, like screensavers and antiviral programs, do not need to be shut off, but this is not a universal rule. If the new program's documentation tells the user to shut off such programs, it is a good idea to do so. Some programs will simply not install correctly without this step.

Application Upgrades

The evolution of a piece of software often continues after it first comes on the market, and newer versions are made available as *upgrades.* Upgrades are updates, which are not stand-alone programs, that are only meant to add new features to an existing program. For example, one version of a Web browser may not be capable of reading a certain type of file, but a newer version could incorporate that ability. Rather than forcing a user to buy the newer version at its full price, the software makers often offer the upgrade at a reduced price. It is installed by the user, transforming the older version into the newer.

There is an important caveat about upgrades and backward compatibility. Generally, the newer version of an application will read files from the old version, but the reverse is not necessarily true. Files from the newer version may be unreadable by the old version. This is especially true of applications, like word processors, which actually create their own documents.

Examining system requirements and installing an upgrade is much like installing any other software. If basic steps are followed, installation will generally proceed smoothly.

Utilities

Utilities can perform a variety of functions to help the user manage increasingly complex computers. They range from disk defragmenters to enhanced file managers to programs that will give the user more control over desktop icons. If you can think of a problem, or a procedure that is cumbersome or difficult to perform, there is good chance that someone has written a utility to address the problem.

Disk Utilities

Disks play a vital role in computing, and they are also one of the most fallible elements of any system. For this reason, *disk utilities* should be considered a necessary addition to every system. A disk utility can diagnose and repair a variety of disk problems. In some cases, corrupted data can be recovered with a utility when it might otherwise be lost. In general, disk utilities take features, like Defrag or Scandisk, built into some operating systems and raise them to a higher level of effectiveness. A utility that is used even once to recover valuable data can provide justification for its purchase in a matter of minutes.

System Utilities

System utilities and disk utilities are closely related, and in fact, they may come as part of a utilities suite. But system utilities go beyond diagnosing disk problems. They can perform a number of other functions, such as diagnosing conflicts between hardware or software elements, and they can even recommend specific maintenance steps to speed a computer's performance. System utilities can also offer less directly useful, but nonetheless interesting, functions, such as indicating how much of a computer's RAM is in use at a given moment, or what proportion of a CPU's possible workload is being used. Sometimes these features are displayed on a virtual "dashboard" to allow the user to monitor system performance continuously.

The software used to diagnose extension conflicts in Apples falls into this category. Because of these extension conflicts, and the lack of maintenance utilities in the Mac OS, system utilities are particularly vital for Apple computers.

Drivers

Drivers play an often invisible, but nonetheless vital, role in computing. Without them, a computer would largely be unable to communicate with hardware devices like drives or printers. If the system is functioning correctly, after installation a driver may go on silently doing its work in the background for the rest of the computer's life span.

Drivers bear a close relationship with the computer's BIOS. The BIOS, Basic Input and Output System, is imprinted on a ROM chip and takes care of communication between hardware devices. However, there are far too many possible hardware devices for all of the necessary commands to be stored in a reasonable amount of memory. Drivers are used instead and are used as extensions of the BIOS by the computer. When, for example, the user sends a document to be printed, the driver and BIOS take care of the low-level system operations that actually accomplish the printing.[2]

Like any other software, drivers may be updated periodically. This happens less frequently than with application and utility software, but it does occur. If drivers were etched permanently into ROM with the rest of the BIOS, the BIOS would be obsolete almost instantly as various drivers were updated.[3]

Users can frequently ignore driver updates if they are satisfied with the performance of a hardware device. An updated driver may simply add features or small amounts of speed that are inconsequential to a particular user. There are several exceptions to this, though. If a machine's operating system is upgraded, the drivers used with the former operating system may not work. Most will, but not all. If a hardware device refuses to work after an OS upgrade, the solution may be to install a new driver. Updated drivers are often available free of charge from a manufacturer's Web site, along with instructions on how to install them.

The other exception occurs when the driver requires a *patch.* A patch is similar to an upgrade, but it is generally smaller in scale and is meant to address a specific flaw in a piece of software. For example, a sound card's driver might malfunction under certain conditions that are discovered only after the card has been used in the field by consumers. Rather than allowing users to return the cards, the maker could make a patch available to solve the problem. Patches are often free, and they are also used to solve problems with virtually any type of software, not just drivers.

Like any other piece of software, drivers can become corrupted. If a piece of hardware is malfunctioning in a seemingly random manner, or not working at all, reinstalling its driver can be a good troubleshooting and repair step.

Applets and Plug-ins

The surge in popularity of graphics applications and the World Wide Web has created high demand for small pieces of add-on software called *applets* and *plug-ins.* They perform similar roles but do so differently.

An applet, sometimes called a *helper application,* is a small application that performs a highly specific function, often as an adjunct to a larger application. A common example would be one of the large number of games that can be played over the Internet. They are often written in the Java programming language, and they are meant to be played from within a browser like Netscape or Internet Explorer. When the game applet is downloaded, rather than appearing in its own window, it will appear within the browser's window.

The term *applet* can also refer to a small program that runs independently but has limited functions. As an example, Windows includes a small calculator application in the Accessories program group. This application takes up a small amount of disk space and has the basic features one would expect from a calculator. However, it lacks advanced features, such as graphing. The use of an applet in this sense is less common.

Plug-ins, by definition, are always used as part of a parent application. They give the application to which they are attached new features or enhanced operations.[4] One of the more common uses of plug-ins is with graphics programs. The graphics plug-ins frequently come in the form of *filters,* which create special effects or controlled distortions of graphics files. There are also plug-ins for browsers, which allow the browser to read files, such as explorable 3-D worlds, that the browser normally would not be able to process.

Plug-ins or applets are frequently free on the Web, and they are often designed to use files created by entirely separate applications that are sold by the manufacturer. Perhaps the best-known example of this is the Acrobat Reader from Adobe. Acrobat is an application that creates files that retain the look and feel of a printed page, complete with fonts and graphics. It is often used to produce Web equivalents to printed publications. The Acrobat reader can be used as a helper application with a browser or as a stand-alone application. The reader software allows the user to download and read Acrobat files but does not allow the actual creation of Acrobat files. Making the reader available free of charge is seen as a way to encourage the use of the Acrobat application, which is not free.

We can expect applets and plug-ins to continue gaining popularity as more people learn Java and as more information is distributed over the World Wide Web.

Summary

Software turns a computer from a simple computing device into a useful machine. Application programs provide the widest variety of capabilities. Productivity applications often come as a suite and have great general utility. Cataloging software is the most specifically related to library functions, and a cataloging workstation with access both to CD-ROMs and network resources can streamline workload.

OPACs have become the heart of many libraries and can be administered by library staff or information technology administrators, depending on the skill level of the staff members involved.

Utility software can maintain and improve a computer's performance, and should be installed on every system. Applets and plug-ins add functionality to their parent applications and will become more popular along with the rise in use of graphics programs and Web browsers.

Glossary Terms

app
applet
application
bundle
cataloging workstation
client
client/server model
disk utility
dumb terminal
filter
helper application
HTML
Hypertext Markup Language
IDE
integrated development environment
interactive development environment
license
multi-user license
office applications
patch
plug-in
productivity applications
single-user license
site license
suite
system requirements
system utilities
upgrade
WYSIWYG

III

NETWORKS AND SECURITY

12

Networks

Networking stands as one of the broadest topics within computing, if for no other reason than the almost infinite number of ways in which a network can be built. Networks range in size from several computers sharing a printer to the expanding vastness of the Internet. While it is far beyond the scope of this book to conduct a definitive discussion of networking, a book like this can give a basic taxonomy of networks and outline the basics of what is involved in implementation and use of each type.

Types of Networks

Local Area Networks (LANs)

Local area networks, usually called *LANs,* are networks that span a small physical area. This can be a single room in which a small number of computers share a printer, a campus-wide network at a college, or a corporate network covering a small number of offices or buildings.

Within libraries, simple LANs can increase efficiency by allowing *shared peripherals.* A shared peripheral is any hardware device, such as a printer or CD-ROM drive, which is used by more than one computer. In the case of printers, for example, it allows multiple computers to use a single printer, allowing the economy of only buying one type of paper, toner cartridge, and so on, and simplifying maintenance.

Shared CD-ROM drives are perhaps one of the most useful of shared peripherals, and they are often shared in the form of a *CD-ROM tower.* Such a tower allows multiple CD-ROM drives to be stacked or mounted in the same housing, often without a separate CPU. For each computer that connects to the tower, the network is transparent, and the computer can access the shared drives as easily as it can access its own.

CD-ROM towers act as a shared peripheral only when they contain small numbers of drives and when they are connected to a relatively small number of computers. If a large number of drives are to be connected, or they are to be accessed by many machines on a large LAN, it is more effective for them to be attached to a server connected to the larger network. This arrangement offers benefits of scale but requires a significantly greater amount of time to set up and administer. The largest towers and servers may require a full-time administrator.

LANs are by no means restricted to such shared peripherals, though. A larger LAN, such as on a college campus, may offer services, such as e-mail, and may house large databases of academic and administrative information. It is not unusual for colleges to actually have several LANs. One might hold administrative information, such as payroll or course enrollment lists, and access to that smaller network could be tightly controlled. Another LAN might be devoted to OPAC use and yet another to storage of Web-based information. Sometimes these are separate networks, and sometimes they are simply separate servers combined into a *Campus Wide Information System,* or *CWIS.* A CWIS allows users varying levels of access to the servers on the network.

A newer concept, related to LANs, is the idea of an *intranet.* An intranet is a network that functions just like the Internet, using browsers, like Netscape or Internet Explorer, to access information on the network. The crucial difference is that an intranet is not accessible to users outside the organization. Intranets offer a secure means of transmitting information in a browser-accessible format that is comfortable for an increasing number of users. Intranets almost invariably require large amounts of administrative time, and they are most often used by private companies rather than academic, school, or public libraries, though this may change as networking technology becomes increasingly commonplace.

Metropolitan Area Networks (MANs)

Metropolitan Area Networks, referred to as *MANs,* are medium-size networks that serve an area the size of a city. They are closely related to *Freenets,* which are volunteer-staffed and -funded networks that are often compared to public libraries.[1]

MANs take a variety of forms, and many are not free. They are networks like any other and may require membership. The crucial difference is that they frequently offer information tailored to residents of a specific community. In fact, a metropolitan network can be a good source of local information for a library.

Freenets, because they are noncommercial, tend to offer less to users than other metropolitan networks. They may be as simple as a *bulletin board,* a single computer, or small group of computers to which a user can dial up to download information or programs. Some bulletin boards may offer basic e-mail service as well.

MANs and Freenets are not as common as LANs, and they are less accessible. For these reasons, they are usually of limited use to libraries.

Wide Area Networks (WANs)

Wide Area Networks, or *WANs,* cover large geographic areas. The most frequently cited example of a WAN is the Internet and its global reach. However, properly speaking, the Internet is actually a network of networks. There are a large number of WANs, which each cover a region of a country, state, or other large area, and together they make up the Internet. But for the purposes of the ordinary user, the Internet behaves as one large WAN.

Wide area networks are becoming one of the most important resources libraries possess. The Internet is outgrowing its image as a messy, crowded place where only experts can find quality information, or a place where only an elite few are capable of creating meaningful sites. An increasing number of journals and databases now appears only on the Internet, and libraries are beginning to look seriously at acquiring parts of their collection in electronic form rather than print. For libraries with a strong technological infrastructure, this is a very viable option. But for institutions with a weak or slow connection to the Internet, print may remain the preferable option for the near future.

Libraries are also becoming more active participants on the Internet. Many library homepages have gone from being repositories of simple information like library hours and policies, to being highly developed finding aids that contribute to the utility of Internet-based information. An increasing number of librarians feels comfortable creating and maintaining Web pages, and they are finding, classifying, and speeding access to a variety of resources.

Hardware and Software Requirements

Hardware requirements for networks vary greatly according to the type of network being constructed. A simple LAN may require no more than several extra cables and some basic software knowledge, while WANs require large, complex combinations of hardware and software in a system more similar to the phone company's than to what one would expect in most libraries.

For most purposes, library network needs fall into two broad categories: construction of small, intralibrary LANs for the sharing of hardware devices and information between PCs; and connection to the Internet in one form or another. While there are certainly more types of arrangements than these, together they account for a large percentage of library networking. Construction of WANs is not only beyond the scope of this book, but it is also very specialized work requiring a different set of skills than most librarians will want to learn.

LANs

The most common type of LAN in libraries involves the OPAC server and the machines that access its information. In medium to large libraries, the server software most often will be either UNIX or VMS. VMS is a proprietary operating system used on machines made by Digital Equipment Corporation. It is somewhat similar to UNIX in being primarily a network operating system, but it has never had the popularity or prestige of UNIX. There is also a version of UNIX, named Linux, which is meant to run on PCs and could be a possibility for smaller OPAC networks, which do not require a particularly large server. Linux is a fully functional version of UNIX. However, it has all the complexities of UNIX as well as the features, and smaller institutions that use a PC as their server may prefer to use Windows NT.

The choice of operating system for the server is often dictated by the choice of software. Cost, usability, the number of bibliographic records, and the number of concurrent users may significantly narrow a list of possible OPAC software, and many vendors gear their software for a particular OS.

Libraries need to be aware that UNIX, VMS, and, to a somewhat lesser extent, Windows NT, require significant administrative overhead. It may be a wiser decision to let the information technology (IT) department at a larger institution take care of these chores. IT departments routinely administer a number of servers, and it is often more economical and practical to have them administer an OPAC server as well. This has the advantage of incorporating already existing expertise, rather than training librarians whose primary skills are not in network administration.

In many public and school libraries, which are often freestanding and not connected to an external network, this sort of arrangement may not be practical. In such institutions, an economic decision needs to be made to either train a librarian to administer an OPAC server or hire a specialist whose background is in computing rather than librarianship. As a general rule, larger collections require more complex software and more robust servers.

For many libraries, though, OPACs are no longer their main networking issue. OPACs are fast becoming standard technology with well established requirements. The more common type of LAN libraries want to construct is for shared CD-ROMs, printers, and files. These networks often have very different needs than OPAC networks. While OPAC software most often combines networking and cataloging functions, separate software is needed to build LANs for sharing resources like printers and CD-ROM drives. This software is usually referred to as a *network operating system.* A network OS is different from a normal OS in that it is not used to run applications on each individual computer. Rather, it may only provide a command structure for computers to use in communicating with each other. While the individual computers on the network are running non-network applications, a normal OS, like Windows 95, is used. The most notable exceptions to this are UNIX and Windows NT Server. Both UNIX and NT Server can act as an OS for individual machines and also as a network OS that can be used for large and small networks.

Other network operating systems, such as LANtastic and Novell's Netware, are used almost exclusively for their networking capabilities. Netware is probably the more common of the two, and it is used in addition to the computer's normal operating system, which can be DOS, Windows, the Mac OS, OS/2, or UNIX.[2] LANtastic is actually a family of software, with different versions geared for different network applications.[3]

Apples come with built-in networking capabilities in the form of a network protocol named Appletalk, which an Apple can use to communicate with laser printers and with other Apples.[4]

Small LANs that lack a central server are referred to as peer-to-peer networks. They are the simplest to construct and allow the computers to equally share drives, printers, and other peripherals. Larger LANs, with more powerful machines used as servers, are sometimes referred to as "hierarchical," because the server exerts more control over the network than the smaller peer computers.

Hardware requirements for networks vary at least as widely as software requirements. The most simple networks, between just two computers, require nothing more than a cable to connect the two machines, usually through their serial ports. Laplink is perhaps the most well-known piece of commercial software for this, and recent versions of DOS have a built-in feature named INTERLINK, which also allows information transfer between two computers.[5] These connections are easy to set up, but they are among the slowest of possible connections.

Most networks, even fairly small LANs, require a *network interface card,* sometimes called a *network adapter card.* These are expansion cards that allow a computer to communicate effectively and quickly with a network. Most small networks use a standard called ethernet, and ethernet cards are available for a variety of expansion slot types. Ethernet dictates what kind of cables should be used, how machines will communicate over those cables, and a number of other things about a network's setup, or *topology.*

The details of cabling and setup are one of the areas beyond the scope of this book. But one rule of thumb needs to be stressed. The software, hardware, and cabling required will be different for different types of networks. It is not uncommon to hear stories of someone purchasing all the equipment for a network, only to discover that their very expensive cables did not fit the ports on the network cards. Even a small LAN is a significant undertaking, and careful planning is an absolute necessity. Outside help may also be needed.

Internet Connections

There are two main ways of connecting to the Internet: a *dial-up connection* and a connection through a LAN. Dial-up connections are made with a modem via normal phone lines. For libraries in larger institutions, which are connected to a medium or large sized LAN, the local network may be connected to the Internet already, and the network's administrator can tell individual users how to use their Internet access.

Dial-up connections can be tricky to set up, but, once they are running, they tend to need little or no maintenance. A modem and *dialer software* are the initial elements needed to establish the connection. Dialer software can be called various things, including a *winsock, TCP/IP software,* just a *dialer,* or various similar names. Under whatever name they are given, these small applications perform the same function. They allow the user's computer to connect to a server over normal phone lines and to communicate with that server using *TCP/IP.* TCP/IP stands for Transmission Control Protocol/Internet Protocol, which together form the standard for how information is passed over the Internet. Any computer connecting to the Internet must have software capable of understanding this standard. To use a simple analogy, the dialer software is like a pipe coming into a house to connect it to an outside water supply.

If the dialer software is a pipe carrying water, a *browser* is like a showerhead. Rather than undifferentiated information out of the virtual pipe, the browser formats the information in a usable form. On the Web, graphics files, text, sound, and so on, are all downloaded separately, then reassembled by the browser into readable form as a Web page.

Those are the basic elements on the user's end for establishing a dial-up connection. However, users with this type of connection cannot simply dial into the Internet. They must have an *account* with an *Internet service provider,* or *ISP.* An ISP is a company that rents space on its servers to users, and this space is referred to as the user's account. Along with this space comes Internet access and possibly other services like e-mail, file transfer, and access to news groups. Once the connection to the server is established, the server acts as a conduit for information coming from the Internet to the user's machine. A good ISP will provide instructions and user support to aid with setup.

One fact about dial-up connections should always be remembered. Because the link from the user's machine to the ISP is over a normal phone line with a modem, dial-up connections are virtually always slower than the direct connections available to users of the LANs or WANs, found at large institutions. Modems come in various configurations, and their speed is increasing, but they are still, as a rule, relatively slow connections. The various types of modems are discussed in a separate section of this chapter.

An account with an ISP will often include the ability to mount a homepage on the World Wide Web. Most accounts now include a certain amount of disk space on the server, usually anywhere from 1 to 10 MB, which the user can use for file storage, saving e-mail, mounting Web pages, and so on. They may also give the user access to a UNIX shell.

There are typically a number of possible pricing structures an ISP may use. The simplest structures charge the user a flat rate for unlimited access. Users can spend as much time online as they want, with no extra charge. They can also use their disk space allotment as they choose. The flat rates generally start at approximately $10 per month.

Many ISPs now use a modified flat rate. The basic rate will include a certain number of free hours of online time per month, often between 50 and 100. If users exceed this number, they are charged a per-hour fee after that, with the fees varying widely between ISPs. Likewise, if users exceed their disk quota, they will be charged for each additional unit of storage, usually measured in kilobytes or megabytes.

It is vital to note that in this context the library is the user, not individual employees or patrons. Each individual patron may only use a small amount of time, but all the user time added together will be charged to the library's total usage. Given the number of hours most libraries are open, a basic quota of online hours can be used up *very* quickly. In addition, an ISP may require that each individual machine in a library have a separate account.

ISP service may entail other hidden costs if a library chooses to put up its own homepage. Some ISPs charge based on the amount of activity a page generates. If a library puts up a busy page, there may be additional charges based on the number of times the page was visited or the amount of data that was downloaded by users. A library putting up a homepage through an ISP needs to look very carefully at the ISPs pricing structure to avoid surprises.

Variable pricing may sound complicated, but it can have advantages. Unlimited usage generally leads to a very busy ISP, with users who have no reason to limit their online time. The ISP may then struggle to keep up with the demand on its servers and modem pools. A cheap rate is no bargain if it only buys a busy signal. Some of the more expensive ISPs are a better bargain, because the extra money buys better capacity and faster service. Word of mouth is often the best way to sort out local ISPs, so talk to friends or colleagues who have dial-up connections before making any commitment.

Use of an ISP applies mainly to freestanding libraries, such as public or school libraries, which are not attached to a larger network. But for academic or corporate libraries, or public and school libraries with access to a city network, it is almost always easier and cheaper to connect to the Internet through a local network. Once the network is built, the hardware and software requirements for each machine on the network are usually quite straightforward. A network expansion card and network software take the place of the modem and dialer software, and most of the configuration will probably be done by the network administrator.

Some freestanding libraries have been successful in establishing connections with local college networks. Rather than dialing in over normal phone lines, these libraries have leased lines that are always available and may provide much faster access than normal phone lines. Local colleges, and libraries with these connections, would be good places to check into the feasibility of this sort of arrangement.

Modems

For libraries without a direct connection, a *modem* is a necessity for connecting to the Internet. *Modem* is short for modulator/demodulator. A modem converts a computer's digital signals to tones for transmission over a phone line, which cannot transmit normal digital signals. When receiving, a modem does the opposite. It demodulates the tones and turns them into digital data readable by the computer.

There are a variety of modems available, but the choice between them is generally fairly simple, with two basic considerations: the modem's speed, and whether it mounts internally as an expansion card or externally as a desktop component.

Internal modems are generally slightly less expensive and somewhat faster than an external modem. Internal modems usually plug into a standard 16-bit ISA slot, or, in laptops, a PCMCIA slot. The drawback is that internal modems come with all the attendant complications of installing and configuring an expansion card. External modems connect to the computer via a serial port and are generally easier to configure. The choice between different types may depend on the user's comfort with installing an internal modem, on price, and on the availability of an expansion slot or serial port on the computer in question.

When it comes to modem speed, there is really only one consideration. A user should obtain the fastest modem possible. Modem speeds are virtually always slower than a direct network connection, and speed is vital. This will be more true as large multimedia and graphics files become more common on the Internet.

Modem speed is measured in kilobits per second, or *kbps,* which measures the number of bits of data a modem can send or receive each second. Currently the most common speeds are 28.8 and 33.6, or 28,800 bits per second and 33,600 bits per second, respectively. Slower speeds, like 14.4 (14,400 bps) and 9,600, are too slow for current Internet use. Existing modems running at those slower speeds should be considered candidates for replacement, and newly purchased modems should run at nothing less than 28.8. Any good ISP will support these faster modem speeds.

Some newer modems support speeds of 56,000 bps, and they are referred to as *56k modems.* They use different software standards than other modems, though, and there are actually two competing standards: one is referred to as X2, and the other referred to as 56KFlex. Not all ISPs support 56k modems, and an individual ISP may only allow the use of one standard or the other, although a single, combined standard is slowly emerging. For users with ISPs who support them, 56k modems are a faster alternative to 28.8 and 33.6 modems. But never assume a given ISP will support 56k modems. Always check before buying the modem.

Other alternatives to conventional modems are becoming available. The most common alternative is an ISDN modem, which is capable of speeds up to 128 kbps.[6] ISDN stands for Integrated Services Digital Network, and these modems bypass the modulation and demodulation steps used in normal modems. For this reason, they are not technically modems at all. They are more properly called network adapter cards. But because they use phone lines like modems, rather than the direct connection typical of network cards, they are generally referred to as modems.

ISDN modems require special phone lines, so ISDN service is not available in all locations. A library wanting to use an ISDN modem would need to check that its phone connection would support ISDN, and the library would also need to make sure its ISP would also support ISDN. There are generally additional charges involved with ISDN lines, including setup charges and per-minute usage fees. More than other modem types, an ISDN modem requires research before being installed.

While ISDN has a small but noticeable following, another communications protocol is being developed which may overtake ISDN. *Asymmetric Digital Subscriber Line,* or *ADSL,* offers the potential for high-speed communications over normal phone lines.[7] ADSL has only recently begun to receive mainstream attention, however, and it will be a while before the marketplace reaches a verdict on this new technology.

A final alternative, the *cable modem,* is still largely under development and is not widely available. A cable modem allows the transmission of Internet data over the same cable that carries cable television signals. These modems can offer speeds over 500 kbps.[8] It is unclear so far whether they will catch on widely.

Summary

There are three basic types of networks: local area networks (LANs), metropolitan area networks (MANs), and wide area networks (WANs). LANs and the Internet (a WAN) are of the most interest to libraries.

LANs allow efficient sharing of information and hardware devices between computers in relatively close proximity. Commonly, libraries will use LANs either to share printers or to share data on CD-ROMs. LANs can be very simple or very complicated in construction, depending on the type of network being built and the number of machines attached to it. The simplest LANs are referred to as peer-to-peer networks, and they have no central server. Networks with a central server are often referred to as hierarchical networks because the server has more control over the network than the other computers.

Internet connections can be made either through a LAN, which is itself connected to the Internet, or through an account with an ISP (Internet Service Provider). ISPs offer accounts that provide Internet access, file storage, and may allow users to mount their own Web page. Pricing structures vary widely between ISPs, and the alternatives should be weighed carefully before starting an account. The basic requirements for connection through an ISP are a modem, dialer software, and, most typically, a Web browser, though other software, such as file transfer programs, may also be used.

Modems come in a variety of types. The fastest are internal modems running at 56 kbps. Other types of modems are also available, including ISDN and cable modems, as well as ADSL, but these other types enjoy less support by ISPs and phone companies, and their future is uncertain.

Glossary Terms

56k modem
account
ADSL
asymmetric digital subscriber line
browser
bulletin board
cable modem
campus wide information system
CD-ROM tower
CWIS
dialer
dialer software
dial-up connection
Freenet
Internet service provider
intranet
ISP
kbps
LAN
local area network
MAN
metropolitan area network
modem
network adapter card
network interface card
network operating system
TCP/IP
TCP/IP software
topology
WAN
wide area network
winsock

13

Computer Viruses

A *virus* is a small program intended to do harm to a user's machine, usually by deleting data or damaging key system files. These programs are built with that express purpose in mind, and they represent the digital equivalent of vandalism. They can make their way into a computer by a number of means, but the most common way is through the Internet. Any program downloaded from a network can harbor a virus that will begin to do its damage when the program is run. Some viruses begin to act immediately, and others may wait for some specific event to actually do any damage. One prominent example is a virus that remained quietly in residence on users' hard drives until the anniversary of Leonardo da Vinci's birth.

A taxonomy of viruses would be extremely lengthy. Some viruses are extremely sophisticated, reflecting the programmer's high degree of technical knowledge. Other viruses are very good at hiding their own presence until after their damage is done. A full explanation of how viruses work is far beyond the scope of this book. The virus discussion here will focus on how to minimize the chance of infection and how to clean up the damage if one does occur.

Antivirus Software

Antivirus software provides the most potent weapon in combating viruses. Antivirus software attempts to detect viruses, remove them, and repair any damage, if possible. This software is relatively cheap and fairly effective, though not foolproof.

Antivirus software uses two basic methods to do its work. One method is to detect any activity that could indicate the presence of a virus, such as attempts to delete or alter certain system files. The other is to look for *virus signatures,* or *signatures* for short. Virus signatures are unique pieces of the code used to write the virus, and they are often compared to DNA or fingerprints. Once a virus is found, the companies that manufacture antivirus software will examine the virus, find some unique part of its code that constitutes the virus's signature, and add that signature to their library. A vital part of an antivirus program is a database of these signatures. The antivirus software acts by scanning the data stored in a computer's disk and in its memory. If any of the virus signatures are detected, the user is alerted, and the antivirus program will generally have features for getting rid of the virus. These databases can generally be updated free of charge by downloading a new version of the signature database from the software company's Web site. Subscription services that deliver the new signatures on disk are also available.

It is important to note that an antivirus program will lose its effectiveness against new viruses unless it is updated. There is also no guarantee that all possible viruses will be included in the database. In fact, the opposite is true. Because new viruses are constantly being invented, there is a virtual guarantee that some viruses will *not* have been detected, analyzed, and added to a particular signature update. Antivirus programs can drastically reduce the chance of a virus damaging a computer, but they do not offer a guarantee of safety.

Emergency Disks

Because antivirus programs cannot offer a guarantee against infection, they will often provide a way to make *emergency disks.* Emergency disks are floppy diskettes that contain the basic system files the computer needs to start. If a virus has disabled the computer, the emergency disks allow the user to restart the computer and begin the process of reinstalling software files damaged or deleted by the virus. Often other utility packages, ones not related to virus protection, also have a feature for creating emergency disks. These disks are extremely useful in a variety of circumstances, such as with hard drive problems affecting the files a computer needs to start properly. If the copies of these files on the hard drive are damaged or deleted by any means, virus or otherwise, an emergency disk, or disks, can be a lifesaver.

Diagnosing a Virus Problem

Diagnosing a virus that has remained undetected by antivirus software is difficult. Viruses often have subtle effects on a computer, and the user may not find the effects easy to recognize. It is also easy to blame other problems on viruses. Computers have become very complex, with an array of hardware and software elements interacting in an environment where user mistakes can occur. Faulty software, hardware conflicts, and user error are far more likely to cause trouble than a virus. But there are some signs to watch for if viruses are a concern:

- Files that were not deleted by a user disappear
- Programs that worked well in the past begin to develop problems
- The computer begins to slow down noticeably
- Garbled characters or symbols appear on the screen
- The hard drive is active (its light is flashing, or something similar) for no apparent reason
- Some viruses will also announce themselves, as if to gloat
- Error message indicating a lack of memory can mean that a virus is taking up memory space

All of these signs, though, can be caused by other things. For example, some system utilities work when the computer is idle. If the utility program accesses the hard drive, it may appear that the machine is "thinking for itself," or doing something the user did not ask for. To exacerbate the problem, many computers now come with a variety of utilities that are installed and running when the computer leaves the factory. Without the user even knowing, there can be utilities monitoring various functions on the machine, causing the hard drive and other components to be active even when the computer seems to be idle.

Fortunately, most of the antivirus software companies are more than happy to help users who think they may have a virus. It stands to the companies' advantage to discover and analyze new viruses, and they may be glad to hear from a consumer with an infected machine. But because so many things other than a virus are more likely to cause these symptoms, the user should make a real effort to rule out other causes before contacting one of the manufacturers.

It needs to be stressed that *every* machine in a library should have antivirus software, and the signature database on *every* machine should be updated regularly. Even if a machine is not connected to a network, it can still become infected via floppy disks, CD-ROMs, or any other medium used to deliver programs or data.

General Precautions

Antivirus software is not the only way to guard against viruses. A number of other precautions can, and should, be taken.

One of the main sources of viruses, after the Internet, is *pirated software.* Pirated software is software that has been copied illegally to avoid paying the full price. Software piracy occurs worldwide, and some of the perpetrators are highly skilled marketers of their goods. They produce large numbers of pirated copies, complete with photocopied documentation in some cases, which they can sell for far less than the company that had to spend research and development funds to create the program. Not only are many of these dealers crooks, but they are malicious to boot. Pirated software can be loaded with viruses, giving the pirate the added thrill of harming both the software company and the consumer. In a bizarre moral twist, some pirates who add viruses to programs consider themselves Good Samaritans who are punishing the consumers of illegal software.

For all of these reasons, a library should only buy software from a reputable source. Good vendors, who have been in business for a while and want to stay there, have good reason to sell only legitimately published software that the manufacturer created under controlled conditions. It is much the same principle as only buying food whose packaging has not been tampered with.

Any program downloaded from the Internet should be suspect. There are a large number of sites on the Web that allow a user to download various types of programs, and they are among the more likely ones to carry viruses. Most antivirus programs can scan these programs before they are run, but they may not do so automatically. The user may need to manually scan a downloaded file, and the documentation with the antivirus program should explain how to do this if it is not done automatically.

Another qualification is necessary here. Many very reputable software companies offer software for download from their Web sites. These downloads are usually as safe as buying that company's software off the shelf. The sites to be leery of are the various archives of software from various, often unknown, sources. Many of these archives do no virus scanning before a program is made available to the public, and they offer fertile ground to virus authors looking to cause damage.

Viruses can move through any medium used to carry programs or data, and this includes diskettes. To prevent infection through this means, some institutions prohibit users from bringing their own floppies. This is a drastic measure, though, and can cause substantial inconvenience for users. Unless the library is prepared to sell or give diskettes to users, this step is a questionable one.

Security Software

Not all risks originate from viruses. Malicious users, and users who simply make mistakes, will frequently cause more problems than any virus. This is especially true of public access computers, where users of widely varying skill levels have free use of the machines. Fortunately, there are pieces of software that can protect a library's machines from much of this mischief.

Every public access computer should have software meant to shield its essential system files from tampering and deletion. Two of the more common applications of this type are Winshield, made for IBM-compatibles by Kent-Marsh, and FoolProof, made by SmartStuff Software, for Apples. Programs like these offer varying levels of protection. They can prevent the deletion or moving of icons, the alteration of system files, and their advanced features can prevent users from using any program not authorized by the security software. The extent to which a computer should be restricted with this software is a matter of local policy, but, at the very least, vital system components should be shielded from harm.

Steps can also be taken to monitor the use to which a computer is put by users. Sophisticated monitoring software can keep a log of what applications are used, what Web pages were visited, how many pages were printed, and so on. These monitoring applications may require users to log on so that their individual use can be monitored, or the application may keep more general track of a computer's activity without linking it to specific users.

Monitoring software offers libraries the ability to track computer use in the same way they track circulation. It may, in fact, be the only practical way some libraries have to see which online databases are popular with patrons, and which are not used. If access to a database is bought with a flat fee, the library may not receive the kind of specific usage reports generated by services that charge in proportion to actual use. In that case, monitoring software can aid collection development decisions. However, this software may create extreme privacy issues. Being able to identify a specific user's activity is an easily abused capability, and librarians should use it judiciously.

Libraries also find themselves under increasing pressure to filter out material from the Internet which some patrons might find objectionable. This raises many troubling questions and may actually violate ALA guidelines, which read in part, "Although electronic systems may include distinct property rights and security concerns, such elements may not be employed as a subterfuge to deny users' access to information."[1] But filtering software also offers filtering capabilities that may have nothing to do with restricting content.

Many libraries want to set certain machines aside for specific purposes. This is very common in academic libraries, where computers may be considered research tools, distinct from machines used for e-mail, word processing, and other personal use. Filtering software gives the library the capability to deny access, for example, to sites on the Web that provide free e-mail accounts or recreational chat rooms. This would cut down on frivolous use of the computers, leaving them free for patrons doing research.

Filtering for broad categories of use, rather than for content, still creates ethical issues, though. For some patrons, free accounts on the Web may be their only access to e-mail. Also, research is frequently difficult to define, and many librarians are understandably hesitant to monitor patrons' computer use that closely. One compromise is to provide unrestricted access on some machines, while restricting others to research use. This assures that the full range of Internet activities is available to patrons, without making online research impossible because of crowding at every terminal. As with any policy that has a possibility to restrict access, filtering policy needs to be carefully considered before implementation.

Summary

The free flow of information over networks, and the use of public access computers by untrained patrons, creates security risks for libraries. There are no foolproof ways to eliminate these risks, but they can be reduced.

Viruses, hostile programs that can damage computers, are a common risk on networked computers. Antivirus software, with frequently updated databases of virus signatures, constitute the single best defense against viruses. Every computer in a library should have antivirus software, even if the machine is not connected to a network.

Software should also be used to shield important system software components from tampering, damage, or deletion. This same software can also restrict which applications a patron is able to use.

Monitoring the specific activities for which a computer is used can be a valuable planning tool, but it also raises ethical issues. Any policy that might serve to restrict access or risk a patron's privacy must be implemented very carefully, preferably with reference to established guidelines.

Glossary Terms

antivirus software
emergency disks
pirated software
signature
virus
virus signature

IV

FUTURE DEVELOPMENTS AND PURCHASING

14

Network and Low-Cost Computers

The content of this chapter overlaps with many areas of the book. Network and low-cost computers represent a move away from the relatively complex, stand-alone PCs that this book focuses on. But many of the technical issues around both hardware and software remain the same, and in fact constitute necessary background to fully understand the machines discussed in this chapter. It is unclear what role these newcomers will play, but they at least have the potential to fill certain niche roles for libraries.

Many users and system administrators complain that current computers are too complex and expensive and have features that few users actually need. These systems often seem to require far too much maintenance and, in the case of public access computers in libraries, suffer from too much downtime. For library administrators who watch technology costs rise each year, lower cost, lower maintenance computers can sound like a panacea. While the low-cost computers are a current reality, network computers and other alternatives to traditional PCs are still in a developmental stage and need to be scrutinized even more carefully than other machines before purchase.

Low-Cost Computers

The term *low-cost computer* has entered our technical vocabulary and is no longer just a vague term for an inexpensive machine. It is often used synonymously with the term *sub-$1000 PC*. Not long ago, that sort of usage probably would have referred to an older computer bordering on obsolescence. But in the current computing environment, low-cost computers are the result of a deliberate push by manufacturers to provide inexpensive, relatively simple machines that still provide users with a reasonable level of performance. Intel's Celeron processor, mentioned in the chapter on CPUs, represents an effort to aim hardware directly at this market.

In addition to relatively modest CPUs, low-cost computers tend to have smaller amounts of memory (generally 16 MB), smaller hard drives, and modest graphics cards. These computers are meant for users who run relatively simple applications and do little or no multitasking. In libraries, their most likely application is for use as patron terminals for OPAC searching, use of text-based CD-ROMs and databases, and browsing Web sites that are not resource intensive.

In practice, this may seem no different than deploying an older machine or buying a nearly obsolete machine as a bargain. There are several important differences, though. A low-cost, but new, computer will not have had the wear and tear of an older machine, and it will presumably come with a good warranty. In addition, the newer components, while inexpensive, will probably incorporate some newer features and improvements not necessarily available for outdated models. The benefits and drawbacks of these inexpensive machines will make more sense in the context of the next section on network computers.

Network Computers/Machines

Network computer, in common usage, refers to a number of different machine types, all of which attempt to address the problems associated with rising costs and increasing complexity. The term has a more precise meaning, though, and it is worthwhile to spell out the various types of computers that fall under this general rubric,[1] and what advantages or pitfalls each might hold for libraries. *Network machine* will be the general term used here.

Network Computers

Network computers, variously called *network PCs, NPCs, net PCs,* or *NCs,* bear a superficial similarity to regular PCs. They run normal productivity applications, can be used to browse the Web, and perform many of the other usual activities of computing. The critical difference is that network computers do not store the software themselves. The software resides on a server, and when a network-computer user starts an application, the software is passed over the network and executed locally on the network computer. At first this arrangement may seem cumbersome, but it offers a crucial advantage: the software is installed, configured, and maintained in only one place, rather than on each individual machine. Similarly, any upgrade is done only once, on the server. There is no need to upgrade and reconfigure the software for each computer. There is also no need to worry that library patrons or staff members will accidentally, or intentionally, delete software or change critical settings.

Windows Terminals

Windows terminals are an implementation of *thin client* technology. Thin clients are essentially dumb terminals with graphics capabilities. As with dumb terminals, all significant processing and storage take place on the network server. The only information passed over the network to the terminal is information about what should be displayed on the screen, and the terminal passes user input back to the server. But unlike dumb terminals, which allow only text-based display and input, Windows terminals allow for a graphical interface and sound. As the name implies, Windows terminals run the same Windows-based applications one would expect to use on an IBM-compatible. In libraries, these terminals are probably best used as Web terminals or as staff machines for the use of common applications, like word processing and spreadsheets. In schools, they have a clear role to fill in labs where students use common software running on a server.

Managed PCs

Managed PCs frequently crop up in discussions of network and low-cost computers, but they actually represent a very different computing model. Managed PCs are "fat," in that most or all of their programs and data reside on the PC itself, and not on the network server. In fact, many managed PCs are more robust, and incorporate more features, than a normal PC. The critical difference between managed PCs and normal machines lies in features that allow for administration and maintenance of a managed PC from a remote location. Some of these machines can be "woken up" or turned on by a network administrator working at a different location. Maintenance, upgrades to software, and other functions can then be performed remotely. Some managed PCs are also set up to stop local users from changing configurations or software.

Net PCs

The term *net PC* is often used to describe a managed PC with relatively modest capabilities and strong security features. They combine the management capabilities of managed PCs, the performance of a low-cost PC, and extra security features, like a locking cabinet to prevent theft or alteration of internal components.

Virtually all of these network machines in their various forms will come with a network card or modem. However, many of them do not have floppy-disk drives, and they may be referred to as *diskless workstations.* Diskless workstations will either have a hard drive or will make use of the network server for document storage. Utilities like e-mail or FTP will be used if the documents need to be sent from a diskless workstation to a location not connected to the local network. Some of the machines will have a parallel port for a printer, but they also lend themselves very well to use of a network printer.

Despite their simplicity, some of the network machines still have expansion slots. A network machine may consist only of a monitor and keyboard, with all of the hardware built into the back of the monitor's case. Most of the machines, though, will have a separate case that plays the role of the CPU, and these machines are more likely to have expansion slots to accommodate upgrades.[2]

Overview

The types of network machines outlined above offer various advantages and disadvantages over traditional PCs. The ability to simplify hardware management and software upgrades is perhaps the strongest of these advantages. In addition, these machines, especially the network PCs and Windows terminals, offer a possible hedge against obsolescence. Because they are only running basic applications, or simply displaying the screen information for applications running on a server, they are more likely to remain capable of performing those duties even as the servers to which they are attached must evolve to keep up.

These machines come with important caveats, though. While they are simpler machines than traditional PCs, they are generally less capable. This is especially true of network PCs and Windows terminals. They are meant for low-end applications, and will probably not perform well on more demanding tasks. Additionally, while they make good replacements for dumb terminals, it will not be a simple one-for-one exchange. Network PCs receive their software over a local network, and Windows terminals receive all the information they need to display on the screen. This represents a substantially larger amount of information than the information required by a text-based dumb terminal. There is no guarantee that a library's current servers and network hardware will carry the load adequately. In addition, the price difference between network machines and low-cost PCs has shrunk considerably, calling into question the economic savings originally promised by network machines. It remains to be seen how the marketplace will react to these technologies.

Summary

The low-cost PCs offer all of the traditional flexibility in software choice and customization for individual needs, and they are more easily expanded. In the case of a network failure, traditional PCs can continue performing non-network functions, while network-based machines would be rendered effectively useless. If library staff members depend on network machines for basic productivity software, a network failure takes on larger ramifications than simple loss of OPAC and Internet access, although those are bad enough on their own. But for libraries with robust networks, and technology management resources to match, network machines may offer efficient, practical alternatives, especially in public access roles.

Glossary Terms

diskless workstation
managed PC
NC
network computer
net PC
network PC
NPC
thin client
Windows terminal

15

Going Shopping

Many sections in this book mention the issues involved in purchasing hardware and software and advise consultation with a good vendor or retailer. Unfortunately, phrases like good vendor and reputable retailer carry little specific substance with them. One of the aims of this book has been to give librarians and information professionals enough technical orientation to understand the specifications displayed in catalogs and advertisements, and to know the practical implications of those specifications. But beyond this technical expertise, it is also useful to be aware of vendors whose products and prices are good yardsticks for measuring the market's current state, and more useful yet to know of good sources for objective information that tells the story behind the specifications.

Vendors

There are two types of vendors: vendors who manufacture hardware or software and sell directly to libraries and consumers, and vendors who are resellers or retailers. The two choices, selecting a manufacturer or brand name and selecting the place to purchase it, are closely related. Some brands of computer, like Dell, are available directly by mail order or over the Web, while a brand like Compaq is available primarily through third-party vendors who may sell locally or electronically via the Web or telephone. Companies like Gateway and Apple sell computers both directly to the public and through local retail stores. So to some extent, choice of brand names in hardware can affect the choice of seller. This is less the case with software. While many software vendors sell their products directly by mail order or over the

Web, most software is also available locally or through established mail-order catalogs not directly tied to the manufacturer.

The Web is rapidly becoming the premier, and cheapest, place to find good buying information both from and about manufacturers. An increasing number of manufacturers sells directly over the Internet, making their Web sites excellent sources of information about features and pricing. Four vendors come to mind as leaders both in manufacturing and in Web retailing:

Dell: www.dell.com
Gateway: www.gw2k.com
Micron: www.micronpc.com
Apple: www.apple.com

This is not a comprehensive list of computer companies selling online, but it is a good cross section. Most importantly, these four vendors allow the user to customize a machine online, choosing features and add-ons, and they provide the user with a price for the customized machine. In general, though, there are many exceptions. Buying a brand name computer over the phone or Web is cheaper than buying one at retail, but these Web sites still provide an excellent yardstick with which to judge an approximate price for a given set of capabilities. Small-scale and local computer manufacturers may have lower prices than the major companies, but their service and quality will be more variable, and a decision to buy from such a company should be made with care.

The companies listed above sell aggressively over the Web, but hardware and software from manufacturers is also sold by good mail-order companies that do business both online and by phone. These companies frequently send out paper catalogs, as well. While there are a number of companies in this category, two of them serve as especially good examples. MicroWarehouse (www.warehouse.com) has a Web site that is divided into four sections that reflect the break down of their print catalogs: MicroWarehouse for IBM-compatibles, MacWarehouse for Apples, DataComm Warehouse for networking needs, and SuppliesWarehouse, which sells accessories like carrying cases, furniture, and media.

The other example is PC Connection (www.pcconnection.com). PC Connection is also broken down into three sections that reflect the division of its print catalogs: PC Connection for IBM-compatibles, MacConnection for Apples, and NetCom for networking needs.

It needs to be emphasized that this is not a definitive list of online or mail-order vendors, nor is the selection available from these vendors presented as definitive of the state of the industry. Many small, lesser-known manufacturers produce very fine hardware and software, and their products may or may not be available through major vendors. However, at any one time the selection and prices of major vendors such as the ones listed above are a good approximation of what is available on the market and roughly what prices a buyer can expect. In addition, these vendors often, to varying extents, offer educational discounts that may benefit libraries as well as schools.

Sources of Buying Information

The choice of vendors is often a library management issue rather than a technical issue. The decision may hinge more on level of service and an ongoing relationship than on strict technical merit. But no matter how positive a relationship a library has with its vendors, it is important to have access to objective information about products the library is considering purchasing. Just as the Web has emerged as an excellent source from which to obtain information directly from vendors, it also has a large amount of information in the form of buyers guides and other reviews of both hardware and software.

One of the best Web sites in terms of breadth is Yahoo's section on Computers and the Internet (www.yahoo.com/Computers). The section is organized into a large number of subsections, many of which contain links to reviews of hardware and software or to buyer's guides. As always on the Web, though, an important caveat applies. Many of the pages offering reviews and guidelines have been put up by well-intentioned amateurs whose knowledge varies widely. Just as libraries should play close attention to a work's authority in selecting a print volume, they should take care in assessing the information on any Web site. However, many of the links for Yahoo's Computer and Internet section lead to useful, reliable information.

An increasing number of computing magazines also maintains Web sites with archives of past articles and reviews as well as features not available in their print counterparts. Two of these sites, one for *Byte* magazine and the other for Ziff-Davis, a prominent computer publisher, serve as excellent examples.

Byte's site, at www.byte.com, features an archive of articles from January 1994 forward, though the articles from 1994 do not include graphics and the three most recent issues at any given time are incomplete.[1] In addition to outstanding articles on virtually every area of computing technology, the site also features Bytemarks, a set of test programs that analyzes the performance level of either an IBM-compatible or an Apple. These programs, called *benchmarks,* test various system functions and return a rating, generally on a numeric scale.

Not everyone agrees on the importance that should be attached to these scores, but they are widely used and can give a good, rough estimate of a machine's performance.[2] Reviews of new computers often feature benchmark scores, and while they should not be treated as a definitive statement about a computer's ability, they can still be a useful measure. The Bytemark suite can be downloaded at www.byte.com/bmark/download.htm.

Ziff-Davis's site (www.zdnet.com) also contains a large amount of useful information. Ziff-Davis publishes over a dozen computer-related magazines, and a full list is available at www.zdnet.com/findit/mags.html. The magazines offer various degrees of archiving, and software downloads are also available from the site. ZDNet also provides an excellent Buyer's Guide, which is available at: www.zdnet.com/netbuyer/edit/howtobuy/.

While *Byte* tends to emphasize new-edge developments over practical tips, Ziff-Davis's magazines are often rich in very practical advice on optimizing current hardware and software. However, neither site focuses exclusively on either area.

There are also sites that can help you select an Internet service provider. Many libraries and schools, or at least school districts, maintain their own networks and Information Technology departments, but many others gain Internet access through commercial providers. There are currently two sites, www.thelist.internet.com and www.isps.com, which provide extensive listings of Internet providers along with a profile of the services, rates, and so on, offered by each. Both sites are searchable and are good places to start if you are looking for a commercial provider.

Summary

Good buying decisions require the synthesis of large amounts of information about hardware, software, or both. But knowing the ideal specifications does not necessarily translate into knowledge of what is actually available on the market or available in a given price range. For this kind of real world, practical information, it is extremely useful to be able to look at Web sites, magazines, and vendor catalogs, which provide a reasonable snapshot of what's available.

Glossary Terms

benchmark

Notes

1. CPUs and Motherboards

1. "iCOMP® Index 2.0", www.intel.com/procs/perf/icomp/index.htm, June 6, 1998.

2. Tom R. Halfhill, "Beyond Pentium II," *Byte* 22, no. 12 (December 1997): 80.

3. *The PowerPC G3 Processor,* www.apple.com/powermac/technologies/g3.html, June 4, 1998.

2. Expansion Cards, Slots, and Ports

1. Mark Minasi, *The Complete PC Upgrade and Maintenance Guide, 8th ed.* (San Francisco: Sybex, 1997), 1381.

2. Scott Mueller, *Upgrading and Repairing PCs, 8th ed.* (Indianapolis: Que, 1997), 15–17.

3. Mueller, *Upgrading,* 112.

4. Mueller, *Upgrading,* 115.

5. Minasi, *Complete PC Upgrade,* 130.

6. Denis Howe, *Free On-line Dictionary of Computing* (*FOLDOC*), wombat.doc.ic.ac.uk/foldoc/foldoc.cgi?accelerated+graphics+port, June 4, 1998.

7. Eric Knorr and Seogju Lee, "Hot Graphics PCs: AGP Boosts 3D Power," www.cnet.com/content/reviews/compare/agp/index.html, June 4, 1998.

8. Minasi, *Complete PC Upgrade,* 150.

9. Minasi, *Complete PC Upgrade,* 1065.

10. Howe, *FOLDOC,* wombat.doc.ic.ac.uk/foldoc/foldoc.cgi?small+computer+system+interface.

11. Minasi, *Complete PC Upgrade,* 153–54.

3. Memory

1. Howe, *FOLDOC,* wombat.doc.ic.ac.uk/foldoc/foldoc.cgi?kilobyte, June 4, 1998; wombat.doc.ic.ac.uk/foldoc/foldoc.cgi?megabyte, June 4, 1998. wombat.doc.ic.ac.uk/foldoc/foldoc.cgi?kilobyte, June 4, 1998.

2. Michael J. Zulich, "The Express Bus: 100 MHz and Memory", *Windows Sources,* www.zdnet.com/wsources/content/0698/hw_bus.html.

4. Storage

1. Minasi, *Complete PC Upgrade, 8th ed.,* 764.
2. Minasi, *Complete PC Upgrade,* 1391.
3. Jeremy Judson, *The Macintosh Bible, 6th ed.* (Berkeley, Calif.: Peachpit Press, 1996), 812–14.
4. Minasi, *Complete PC Upgrade,* 766–67.
5. Mueller, *Upgrading, 8th ed.,* 588–89.
6. Greg Harvey, *DOS for Dummies Quick Reference* (Foster City, Calif.: IDG Books Worldwide, 1993), 15–16, 104–106.
7. Harvey, *DOS for Dummies Quick Reference,* 36–38.
8. Richard A. Henle and Boris W. Kuvshinoff, *Desktop Computers in Perspective* (New York: Oxford University Press, 1992), 499–500.
9. Mueller, *Upgrading,* 761.
10. Howe, *FOLDOC,* wombat.doc.ic.ac.uk/foldoc/foldoc.cgi?digital+video+disc, June 4, 1998.
11. "How to Buy a DVD-ROM," *Computer Shopper Netbuyer,* www.zdnet.com/netbuyer/edit/howtobuy/C0000102/dvd_overview.html, June 4, 1998.

6. Printers

1. Howe, "Point," *FOLDOC,* wombat.doc.ic.ac.uk/foldoc/foldoc.cgi?point, June 4, 1998.
2. Ron White, *How Computers Work, Deluxe ed.* (Emeryville, Calif.: Ziff-Davis Press, 1997), 269–71.

7. Sound Cards

1. Mueller, *Upgrading, 8th ed.,* 501.
2. Mueller, *Upgrading,* 491.
3. Mueller, *Upgrading,* 497.

4. Mueller, *Upgrading,* 497.
5. Minasi, *Complete PC Upgrade, 8th ed.,* 982.
6. Mueller, *Upgrading,* 496.
7. Mueller, *Upgrading,* 498.
8. Russell Kay, "Do You Hear What I Say?" *Byte* 23, vol. 1 (January 1998): 115–16.

8. Optical Scanners

1. White, *How Computers Work,* 171.
2. White, *How Computers Work,* 172–73.
3. White, *How Computers Work,* 171.

9. Power Control Devices

1. Minasi, *Complete PC Upgrade,* 423.
2. Mueller, *Upgrading,* 325.
3. Minasi, *Complete PC Upgrade,* 424–25.
4. Mueller, *Upgrading,* 325.
5. Minasi, *Complete PC Upgrade,* 426.
6. Minasi, *Complete PC Upgrade,* 427.

10. Operating Systems

1. White, *How Computers Work,* 19.
2. Howe, "Shell," *FOLDOC,* wombat.doc.ic.ac.uk/foldoc/foldoc.cgi?shell, June 4, 1998.
3. Mueller, *Upgrading,* 921.
4. Howe, *FOLDOC,* wombat.doc.ic.ac.uk/foldoc/foldoc.cgi?cooperative+multitasking, June 4, 1998.
5. Henle and Kuvshinoff, *Desktop Computers,* 353.
6. Howe, *FOLDOC,* wombat.doc.ic.ac.uk/foldoc/foldoc.cgi?preemptive+multitasking, June 4, 1998.
7. Minasi, *Complete PC Upgrade,* 112.
8. Henle and Kuvshinoff, *Desktop Computers,* 326–27.
9. Henle and Kuvshinoff, *Desktop Computers,* 325.
10. Howe, *FOLDOC,* wombat.doc.ic.ac.uk/foldoc/foldoc.cgi?config.sys, June 4, 1998.

11. Kris Jamsa, *Concise Guide to MS-DOS Batch Files* (Redmond, Wash.: Microsoft, 1994), 1.

12. Harvey, *DOS for Dummies Quick Reference,* 38.

13. White, *How Computers Work,* 31.

14. Ed Bott, "Top 10 Reasons to Upgrade to Windows 98 (And a Few Reasons Why You Shouldn't)," *PC Computing,* www.zdnet.com/pccomp/features/excl0598/top10/welcome.html, June 4, 1998.

15. Bott, "Top 10 Reasons", *PC Computing,* www.zdnet.com/pccomp/features/excl0598/top10/wait3.html, June 4, 1998.

16. Howe, *FOLDOC,* wombat.doc.ic.ac.uk/foldoc/foldoc.cgi?Windows+NT, June 4, 1998.

17. Mueller, *Upgrading,* 922.

18. Howe, *FOLDOC,* wombat.doc.ic.ac.uk/foldoc/foldoc.cgi?Macintosh+file+system, June 4, 1998.

19. Howe, *FOLDOC,* wombat.doc.ic.ac.uk/foldoc/foldoc.cgi?os/2, June 4, 1998.

11. Software

1. Howe, *FOLDOC,* wombat.doc.ic.ac.uk/foldoc/foldoc.cgi?interactive+development+environment, June 4, 1998.

2. White, *How Computers Work,* 25.

3. White, *How Computers Work,* 25.

4. Howe, *FOLDOC,* wombat.doc.ic.ac.uk/foldoc/foldoc.cgi?plug-in, June 4, 1998.

12. Networks

1. Howe, "Freenet", *FOLDOC,* wombat.doc.ic.ac.uk/foldoc/foldoc.cgi?freenet, June 4, 1998.

2. Howe, *FOLDOC,* wombat.doc.ic.ac.uk/foldoc/foldoc.cgi?novell+netware, June 4, 1998.

3. Howe, *FOLDOC,* wombat.doc.ic.ac.uk/foldoc/foldoc.cgi?lantastic, June 4, 1998.

4. Jeremy Judson, *The Macintosh Bible,* 906.

5. Doug Lowe, *Networking for Dummies, 2nd ed.* (Foster City, Calif.: IDG Books Worldwide, 1996), 14–15.

6. Mueller, *Upgrading,* 444–45.

7. Howe, *FOLDOC,* wombat.doc.ic.ac.uk/foldoc/foldoc.cgi?adsl, June 4, 1998.

8. Howe, *FOLDOC,* wombat.doc.ic.ac.uk/foldoc/foldoc.cgi?cable+modem, June 4, 1998.

13. Computer Viruses

1. *Access to Electronic Information, Services, and Networks: an Interpretation of the "Library Bill of Rights"*, www.ala.org/oitp/ebilrits.html, June 6, 1998.

14. Network and Low-Cost Computers

1. Lisa L. Sweet, "Are You Ready to Trade In Your PC?" *Internet Computing* 3, no. 6 (June 1998): 53–63.
2. Sweet, "Are You Ready to Trade In Your PC?" 56–57.

15. Going Shopping

1. *Byte,* www.byte.com, May 31, 1998.
2. Howe, "Benchmark", *FOLDOC,* wombat.doc.ic.ac.uk/foldoc/foldoc.cgi?benchmark, June 4, 1998.

Glossary

8-bit sound: Recorded sound represented digitally with eight bits per sample. See also **sampling.**

16-bit sound: Recorded sound represented digitally with sixteen bits per sample. See **sampling.**

56k modem: A **modem** capable of data transfer rates of up to 56 kbps.

accelerated graphics port: Also **AGP.** A high-speed bus, developed by Intel, for faster graphics performance.

account: Rented disk space on a **server** belonging to an **ISP.** An account usually gives a user Internet access and may provide other services such as e-mail and file transfer.

active speakers: Speakers with built-in amplifiers. See also **passive speakers.**

ADSL: See **asymmetric digital subscriber line.**

AGP: See **accelerated graphics port.**

antivirus software: Software that scans for **viruses** and may attempt to remove any viruses and repair damage they might have caused.

app: See **application.**

applet: A small program usually used in conjunction with a larger **application.**

application: A program designed for a specific function or range of functions.

asymmetric digital subscriber line: Also **ADSL.** A network protocol that allows faster **dial-up connections** over normal phone lines than the connections made with normal **modems.**

AT bus: a 16-bit ISA **bus.** Originally developed for the IBM PC/AT computer.

backup power supply: A device that provides a computer with battery power in case of a failure of normal AC power. Also see **standby power supply** and **uninterruptible power supply.**

batch file: A file containing a list of DOS commands to be executed sequentially.

basic input-output system: A set of rudimentary commands that governs how a computer communicates with hardware devices, like disk drives or the keyboard.

benchmark: A program or suite of programs that tests and measures a computer's performance and, typically, rates the performance on a numerical scale.

bidirectional: Capable of both receiving information from the computer and sending it back. Usually used to describe **ports.**

binary: A system for digitally encoding information, with each of the basic units being represented by either a one or a zero. Also see **bit** and **byte.**

BIOS: See **basic input-output system.**

bit: The smallest unit of binary storage, a one or a zero. Also see **byte.**

boot disk: The disk from which the computer loads vital system files during start-up.

brownout: A partial loss of power to a computer.

browser: An **application** that allows the user to view hypermedia information on the World Wide Web.

bubblejet: See **inkjet printer.**

bulletin board: A single computer or small group of computers to which a user can dial up to download information. May also offer simple e-mail services.

bundle: 1. A group of programs that work closely together and have overlapping functions. Also see **suite.** 2. Programs included with a computer are often said to be bundled with the machine.

bus: A circuit allowing one part of a computer to communicate with another part.

bus width: A measure of how much information can travel over a bus at a time, usually measured in bits. The most common are 16- and 32-bit buses.

byte: A group of eight **bits,** with each group representing a single character in storage. The sequence of ones and zeroes in a byte determines which character is represented. Also see **bit.**

cable modem: An extremely fast **modem** capable of data transmission speeds up to 500 kbps. Uses the same lines as cable television.

cache: See **cache RAM.**

cache memory: See **cache RAM.**

cache RAM: Also **cache memory** or **cache.** High-speed memory, usually in small amounts, used to store and deliver information to the CPU faster than **main memory** can.

campus wide information system: A network on a college campus that may integrate several smaller networks.

card edge connector: A set of electrically conductive strips along the edge of an **expansion card,** which allow the card to connect to a **bus.**

case: The outer, usually plastic, shell of a computer that houses the internal components.

cataloging workstation: A high-end computer with access both to cataloging-related CD-ROMs and to networked resources.

CD: A compact disc. A plastic disc with an aluminum coating into which pits are etched with a laser.

CD-R: A CD that can be written to one time by the user, after which it is read-only.

CD-ROM: Compact disc read-only memory. A CD that can only be read, not written to.

CD-ROM tower: A group of **CD-ROM** drives stacked in a single housing so that each drive can be accessed over a network.

central processing unit: The computer's principal processing **chip.**

Centronics port: See **parallel port.**

characters per second: The number of letters a **dot matrix printer** can print per second.

chassis: The inner, metal frame of a computer. Provides electrical grounding and structural support.

chip: A component made of layers of very thin silicon wafers that are etched with electric circuits. A chip can be used for various types of storage and processing.

circuit board: Usually made of fiberglass approximately 1/8″ thick. These are boards of various sizes that hold electric circuits and components.

client: 1. A computer on a network which receives information from a **server**. 2. The software a client computer uses to communicate with the server. Also see **client/server model.**

client/server model: A network model under which a central computer acts as a **server** of information that is accessed and displayed by the **client** machines.

clock cycle: An iteration by the **CPU** of its most basic functions. The number of **cycles per second** is the most commonly accepted measure of a CPU's speed.

COM port: See **serial port.**

command line interface: An **interface** that requires the user to type in commands manually.

command prompt: In a **command line interface,** a set of letters or symbols followed by a cursor, which tells the user the system is ready for input.

communications port: See **serial port.**

consumables: All of the disposable items used when a printer produces a finished page. Usually defined as paper and toner.

conventional memory: Under DOS, the first 640k of **RAM,** in which most programs must run.

cooperative multitasking: A **multitasking** system in which individual programs decide when they need attention from the CPU. See **preemptive multitasking.**

c.p.s.: See **characters per second.**

CPU: See **central processing unit.**

CWIS: See **campus wide information system.**

cycles per second: The number of times per second that a **CPU** can execute its most basic functions, such as a basic arithmetic operation or shifting a piece of data.

cylinder: Two tracks that are directly opposite each other on each side of a magnetic disk.

data fork: The **fork** in which Apple computers keep a file's data. Also see **resource fork.**

Defrag: A utility built into recent versions of MS-DOS, which performs **defragmentation** on a disk.

defragmentation: The placing of scattered pieces of files in adjacent locations on a disk to speed data access time.

density: A reflection of the number of tracks a disk can hold per inch. Higher density and more tracks per inch create increased storage space on the disk. Also see **double density** and **high density.**

dialer: See **dialer software.**

dial-up connection: A connection between two computers that is established over phone lines.

dialer software: Software used to establish a **dial-up connection.**

digital versatile disc: See **DVD.**

digital video disc: See **DVD.**

DIMM: See **dual inline memory module.**

DIP switch: A small switch, usually part of a set of switches. Most often used to change a setting on a **circuit board,** such as the IRQ.

directory: A location on the hard drive or desk top in which files are located for organizational purposes. Often referred to as a **folder.** See also **subdirectory.**

disk utility: A piece of software that performs maintenance on, and diagnoses problems with, computer disks.

diskette: See **floppy disk.**

diskless workstation: A **network computer** or similar machine without a **floppy disk** drive.

DOS computer: A computer running some version of MS-DOS. Sometimes refers to IBM-compatibles in general.

dot matrix printer: A printer that produces characters and graphics with a **print head** that uses the impact of **pins** against a fabric ribbon to place dots on the page. Also referred to as an **impact printer.**

dot pitch: See **dot pitch interval.**

dot pitch interval: The distance between dots of the same color in adjacent **pixels** of a screen display. Measured in fractions of a millimeter.

dots per inch: The number of dots a printer can place on the paper, per inch. Usually used in reference to **laser** or **inkjet printers.** The larger the number of dots per inch, the better the quality of the printout. Also see **dpi.**

double density: A **density** level that allows a 5.25″ **floppy disk** to store 360K, and a 3.5″ disk to store 720K.

double-sided: A double-sided disk can have information stored on both sides.

dpi: 1. Abbreviation of **dot pitch interval.** 2. The number of dots per inch a **laser** or **inkjet** printer is capable of placing on a page.

dual inline memory module: A **memory module** that must be used as half of a pair.

draft mode: A mode in which a printer produces faster, lower-quality printouts that are not meant to have a finished appearance.

driver: See **driver software.**

driver software: Also **driver.** A small program enabling a hardware component, such as a printer or **modem,** to communicate with the rest of the computer.

dumb terminal: A display and keyboard together that does no processing, but only displays information processed by a network **server.**

DVD: Also **digital video disc** or **digital versatile disc.** A high-capacity optical storage disc capable of holding between 4.7 and 17 GB.

EISA: Short for Extended Industry Standard Architecture. A relatively rare, 32-bit version of the **ISA bus.**

embedded system: A computer system meant to run in a very specific environment, such as a stand-alone kiosk or industrial machinery.

emergency disks: Disks that contain important system files that enable a computer to start properly. Used when system files on the hard drive have been damaged by a **virus** or other problem.

expansion bus: A circuit made to accommodate **expansion cards.**

expansion card: A circuit board that plugs into the **expansion bus.**

expansion slot: A plastic and metal housing that allows an **expansion card** to be plugged into the **expansion bus.**

extension: A piece of software meant to extend the capabilities of an Apple computer.

fanfold paper: Printer paper in which the individual sheets are connected end-to-end for continuous feed through the printer.

file format: A scheme for arranging data so that it can be interpreted correctly by software. An example would be a graphics file, which must be in the right format to be recognized as a picture.

filter: Small pieces of software, most commonly used as add-ons to graphics applications, to allow the application to perform special effects or distortions of images.

fire: A **pin** in a **dot matrix printer** is said to fire when it is pushed outward against the ribbon, placing a dot on the page.

Fire Wire: A very high-speed data bus used primarily for advanced video applications.

five-and-a-quarter: A **floppy disk** five-and-a-quarter inches wide.

fixed disk: A disk that cannot be removed from a computer. Typically a **hard disk drive.**

flash memory: Any of a number of types of **ROM** that can be rewritten under limited circumstances.

floating point unit: A **microprocessor** that performs certain mathematical functions for the **CPU** in order to speed the execution of those functions.

floppy disk: Also **floppy.** A small, flexible disk protected by a relatively hard outer cover.

folder: Another name for a **directory.** So called because many **GUIs** represent directories with a folder icon.

font: A **typeface** or graphic style of text.

fork: Under the Mac OS, one of two sections of a file, each of which holds a different type of information. Also see **data fork** and **resource fork.**

format: 1. The arrangement of tracks, sectors, and clusters on a disk so that the disk can be read by a certain type of computer. 2. A command built into MS-DOS and the Mac OS that formats a disk appropriately for that type of computer. 3. **File format.**

f.p.u.: See **floating point unit.**

Freenet: A volunteer-staffed and -funded network.

frequency response: The range of sounds a **sound card** is capable of producing.

generic/text-only printer: A printer, usually dot matrix, that prints text with a minimum of formatting and no graphics.

graphical user interface: Also **GUI.** A **user interface** that allows the use of icons and dialog boxes to enter system commands instead of typing the commands in by hand.

GUI: See **graphical user interface.**

hard disk drive: A sealed disk drive containing disks made from a hard alloy. Usually cannot be removed from the computer.

hard drive: See **hard disk drive.**

heads: See **read/write heads.**

helper application: An **applet.**

high density: A **density** level that allows a 5.25″ diskette to store 1.2 MB, and a 3.5″ diskette to store 1.44 MB.

HTML: See **Hypertext Markup Language.**

Hypertext Markup Language: A command language used to format documents for display as Web pages.

IDE: See **integrated development environment.**

IEEE-1394: See **Fire Wire.**

impact printer: A printer that places dots on the page with the impact of **pins** against a typewriter-like ribbon. See also **dot matrix printer.**

init: See **extension.**

ink cartridge: A cartridge that holds the ink or toner in an **inkjet** or **laser printer.**

inkjet printer: A printer that produces characters and graphics by spraying a series of very small dots of ink or toner onto the page.

integrated development environment: A closely integrated set of **applications** for use in writing programs in a given programming language.

interactive development environment: See **integrated development environment.**

interface: A set of commands, visual aids, or other tools to help the user communicate with the computer.

interlaced: Refers to a monitor that refreshes every other line on the screen and then refreshes the rest on a second pass, severely slowing the effective **refresh rate.**

internet service provider: Also **ISP.** A company that provides users with **accounts** on its **servers.** The accounts typically provide Internet access and other related services.

interrupt: A request sent by a hardware device to the **CPU,** asking the CPU to interrupt normal processing to perform a task or provide information.

interrupt request: See **interrupt.**

intranet: A network that works like the Internet, but whose content is only available to users within a company or other organization.

IRQ: See **interrupt.**

ISA: Short for Industry Standard Architecture. A common type of **expansion slot** and **bus.** Comes in 8- and 16-bit versions, of which the 16-bit is by far the more common.

ISP: See **internet service provider.**

jack: An input or output connector for a **sound card** accessory.

jumper: A pair of metal pins with a plastic and metal connector. Acts as a switch, turned on when the connector is in place, off when it is removed.

kbps: Short for kilobits per second. A measure of how much data a modem can transmit or receive in a second.

kernel: The core of an **operating system.**

kilobyte (kb): Approximately one thousand **bytes.**

LAN: See **local area network.**

laser printer: A printer that uses an internal laser to bond toner to the printed page. Produces sharper characters and images than an inkjet or dot matrix printer.

license: A legal agreement dictating the number of machines on which software can be installed or how many users may use the program simultaneously. See also **single-user license, multi-user license,** and **site license.**

line conditioner: See **power conditioner.**

line printer: A printer that prints one line of text at a time. Often a **generic/text-only printer.**

local area network: A network covering a small geographic area, such as a room, a building, or a college campus.

local bus: A high-speed **bus** typically used for **video cards**.

main memory: A computer's **RAM,** as distinguished from **cache memory** or **ROM.**

MAN: See **metropolitan area network.**

managed PC: PCs that can be turned on and maintained from a remote location.

math coprocessor: See **floating point unit.**

MCA bus: Short for Microchannel Architecture. A relatively rare type of **bus** found on IBM PS/2 computers.

media sensor: A sensor in a **floppy disk** drive that detects the **density** level of a disk.

megabyte (MB): One million **bytes.**

memory module: A small **circuit board** onto which **RAM** chips have been soldered.

metropolitan area network: A network covering a city-sized area and often featuring information of interest to area residents.

MHz (megahertz): Millions of cycles per second. A unit used to measure a **CPU's** speed.

microprocessor: A **chip** designed to execute certain types of commands.

MIDI: Short for Musical Instrument Digital Interface. A **port** on a **sound card** that, along with associated hardware and software, allows a computer to use its sound card to communicate with a musical instrument such as a synthesizer.

millions of instructions per second: Also **m.i.p.s.** The total number of basic operations a **CPU** can perform in a second. Distinguished from **cycles per second** by the fact that newer CPUs can execute more than one instruction per cycle.

MMX technology: A **CPU** enhancement intended to speed the processing of **multimedia** programs.

modem: Short for modulator/demodulator. A hardware device that allows a computer to communicate over normal phone lines.

monochrome: A printer capable of printing in only one color, usually black. Can also refer to a monitor that does not have color capability.

motherboard: The principal **circuit board** in a computer, which holds the **CPU** and other important components.

MPEG: Short for Motion Picture Experts Group. A special **file format** for movies. Requires special hardware and software to be played.

multimedia: The use of more than one medium by a program; for example, a **CD-ROM** that incorporates sound, pictures, and text.

multimedia kit: A **sound card, CD-ROM** drive, speakers, and related accessories sold as a group.

multi-user license: A **license** that allows software to be installed on multiple machines and used simultaneously by multiple users. The license is usually for a specific number.

multitasking: The use of more than one **application** at a time.

naming convention: A set of rules that determines what file names an **operating system** will accept.

NC: See **network computer.**

near letter quality: Also **NLQ.** A mode, usually on dot matrix printers, that produces low-speed, relatively high-quality output. Contrast with **draft mode.**

net PC: See **network computer.**

network adapter card: An **expansion card** that allows a computer to be attached to a network.

network computer: A computer that downloads its software from a network **server** before executing it, as opposed to a normal PC, which stores programs locally, usually on a **hard disk drive.**

network interface card: See **network adapter card.**

network PC: See **network computer.**

network operating system: Software that allows computers to share information and hardware over a network.

NLQ: See **near letter quality.**

nozzle: Small outlet, for ink or toner, in the **print head** of an **inkjet printer.** Each nozzle sprays an individual dot onto the page.

NPC: See **network computer.**

OCR: See **optical character recognition.**

office applications: Applications used for common office tasks, such as word processing, creation of spreadsheets, and so on.

operating system: Low-level software that manages system functions, such as allocation of memory and storage, and communication between different pieces of hardware and software.

optical character recognition: The use of a **scanner** and special software to scan text into a computer in editable form.

optical media: Media read by a laser, as opposed to magnetic media, which are usually read with **read/write heads.**

optical scanner: See **scanner.**

optimization: See **defragmentation.**

OS: See **operating system.**

palmtop: A small computer with limited capabilities meant to run scaled-down versions of common **applications.**

parallel port: A **port** capable of transmitting eight **bits,** or one **byte,** simultaneously. Often **unidirectional,** i.e., can only send information out to a device, and cannot send information back to the computer.

passive speakers: Speakers that do not amplify the output of a computer's **sound card**. See also **active speakers.**

patch: A small piece of software, often free, intended to fix a known problem in an **application** or other piece of software.

PC card: See **PCMCIA card.**

PCI: Peripheral Component Interconnect. A high-speed expansion bus used on IBM compatibles and Apples. Features faster speed than the VESA local bus or EISA.

PCMCIA card: A credit card–sized **expansion card** meant to plug into a small slot, the **PCMCIA slot.** Usually used on laptops.

PCMCIA slot: A small **expansion slot,** usually used on laptop computers, which lack room for a full-sized expansion slot.

pen: The equivalent of the print head in a **plotter.** Capable of drawing a continuous line instead of a series of small dots.

per-page cost: The cost in **consumables** of printing a page with a particular printer.

pin: One of a number of small, pin-like pieces of metal in the **print head** of a **dot matrix printer.** Each pin pushes against a typewriter-like ribbon to place an individual dot on the page.

picture element: See **pixel.**

pirated software: Software copied illegally for sale by unauthorized sellers.

pitch: See **dot pitch interval.**

pixel: Short for **picture element.** A pixel is one of the small dots that make up an on-screen image. Each pixel is composed of three smaller dots—one red, one green, and one blue. Different intensities of these colors are blended to produce other colors.

platform: A combination of an **operating system** and a particular type of hardware commonly used together. IBM-compatibles and Apple computers are two examples.

platter: Each individual disk in a **hard disk drive.**

plotter: A printer usually used to produce technical drawings or plans. Capable of drawing continuous lines on the page instead of the series of small dots placed by other printer types.

Plug and Play: A set of software and hardware standards that allow for easy installation and automatic configuration of components.

plug-in: An **applet** used in direct support of a larger **application.**

PnP: Plug and Play.

Point: A measure of the height of a **font.** A point is 1/72 of an inch.

Port: A physical connection to a computer, which allows data exchange. Also see **parallel port** and **serial port.**

power conditioner: A device that suppresses **power surges,** bridges short **brownouts,** and often filters RFI.

PowerPC: A family of **CPUs** developed jointly by Apple Computer, Motorola, and IBM, now used mainly in Apple computers.

power spike: See **power surge.**

power surge: Also **power spike.** A large increase in the amount of electricity going to a computer.

preemptive multitasking: A **multitasking** scheme under which the **operating system** decides which program receives attention from the **CPU**.

print head: The part of a printer that actually places the small dots making up characters and graphics on the paper. See **inkjet printer** and **dot matrix printer.**

productivity application: An **office application.**

QIC: 1. Quarter Inch Committee. 2. A common standard for **tape drives** arrived at by the QIC.

RAM: See **random access memory.**

random access memory: Specialized chips used to store information in electronic form.

read-only: Any device that can be read by the computer but cannot be written to.

read-only memory: Memory **chips** that can be read but not written to.

read/write heads: Physical components of a disk drive which can read information from a disk or write information onto the disk. Present both in **floppy disks** and in **hard disks.**

refresh rate: The number of times per second that the image on a computer monitor is redrawn. Measured in hertz (Hz, cycles per second).

registry: In Windows 95, a database of system information that the **operating system** uses to integrate the computer's different functions.

resource fork: The **fork** in which Apple computers keep data about a file's icon and the system resources the file uses. See **data fork.**

ROM: See **read-only memory.**

root directory: The uppermost **directory** in a hierarchical file system. See also **subdirectory.**

RS-232: An electrical standard governing **serial ports.**

sampling: A digital recording technique that records sound as discrete samples many times per second, rather than recording continuously as with conventional tape recorders.

scalable: A **font** that is scalable can be adjusted to print in a large number of sizes.

scan head: The part of a **scanner** that passes over the scanned media to create the digitized information.

Scandisk: A utility built into recent versions of MS-DOS that checks both the logical file structure on a disk as well as the physical surface of the disk itself.

scanner: Also **optical scanner.** A peripheral used to visually scan and digitize images or text for use by a computer.

screen resolution: A measure of how many **pixels** make up a screen image. A resolution of 640 × 480, for example, would be made up of 480 horizontal rows, each containing 640 pixels.

SCSI: Short for Small Computer System Interface. A standard for **expansion cards** and controllers that allows multiple devices to be linked together in a chain without each device needing a separate **port, expansion card,** or controller.

serial port: Also known as a **COM port, communications port,** or **RS-232** port. Allows bidirectional communication between a computer and a peripheral, one **bit** at a time.

server: A computer whose main function is to provide information for other computers over a network.

shell: A command interpreter that passes user commands to the **operating system.**

signature: See **virus signature.**

SIMM: See **single inline memory module.**

single inline memory module: A **memory module** which can be used singly.

single-user license: A **license** that allows a piece of software to be used by a single user.

site license: A **license** that allows an institution to install and use software on any machine on its site and allows an unlimited number of simultaneous users.

slot: A receptacle into which a **SIMM, DIMM,** or other **circuit board** can be plugged. Allows connection to a **bus** or other circuit.

socket: See **slot.**

sound card: An **expansion card** that enables a computer to produce higher-quality sound than the machine's built-in speaker. Can also allow the recording or **sampling** of sound and the use of software for **voice recognition** or **voice synthesis.**

spike: See **power surge.**

splitter: A device that allows more than one cable or device to be connected to a **port.**

SPS: See **standby power supply.**

standby power supply: A type of **backup power supply** that switches to battery power after failure of normal AC power. See also **uninterruptible power supply.**

subdirectory: A **directory** or **folder** stored within another directory or folder.

subwoofer: An amplifier that boosts extremely low frequency sound waves to provide enhanced bass.

suite: Also **bundle.** A group of related **applications** meant to work closely together. Often refers to **office** or **productivity applications.**

surge: See **power surge.**

surge protector: Also **surge suppressor.** A device that protects a computer from **power surges** by directing the surge to electrical ground.

system requirements: A list of hardware and software elements that a computer needs to run a particular piece of software.

system utilities: Programs, often sold as a **suite** or **bundle,** that perform or enhance system maintenance or troubleshooting procedures.

tape drive: A drive that uses a robust cassette tape for long-term storage of large amounts of data.

TCP/IP: Transmission Control Protocol/Internet Protocol. Two protocols that together make up the standard for the way in which data is transmitted over the Internet.

TCP/IP software: Dial-up software that allows a computer to use **TCP/IP** to send and receive data.

terminate and stay resident: A piece of software that stays in memory even when it is not being used. Contrast with most programs that are emptied from memory when not in use.

terminator: A device attached to the end of a chain of **SCSI** devices.

thin client: Client hardware or software that performs relatively little processing locally, instead depending on a **server.**

three-and-a-half: A **floppy disk** 3.5″ in diameter.

toner cartridge: A cartridge that holds the ink or toner in an **inkjet** or **laser printer.**

topology: The basic way in which a network is arranged using cables, **network interface cards,** and **servers.**

tracks per inch: Also **t.p.i.** The number of tracks in an inch that a disk is capable of having written onto its surface.

transistors: Small parts of the electric circuits etched onto **chips.** Transistors act as on/off switches to represent binary code.

transparent media adapter: An accessory available for some flatbed **scanners.** Provides backlighting of slides and other transparent media.

trap: An **interrupt request** in an Apple computer.

TSR: See **terminate and stay resident.**

TWAIN: Software used with **scanners** to enable the user to scan an image without leaving a normal graphics application.

typeface: See **font.**

undervoltage: See **brownout.**

unidirectional: Capable of transmitting information in only one direction. Used to describe **ports** that can send information to a device but cannot send any information back to the computer.

uninterruptible power supply: A type of **backup power supply** that runs a computer on batteries that are constantly recharging. The batteries continue to supply power if normal AC power fails. See also **standby power supply.**

Universal Serial Bus: A high-speed **bus** and **port** that allow a large number of peripherals to be connected to a single port.

upgrade: A piece of software that does not function by itself but which adds functions to an older **application,** making the older application equivalent to a newer one.

UPS: See **uninterruptible power supply.**

USB: See **Universal Serial Bus.**

user interface: See **interface.**

VESA bus: A type of **expansion card,** which is almost always a **video card** and is made to plug into the **local bus.** VESA is short for Video Electronics Standards Association, the organization that developed the standard for the card.

video accelerator: See **video card.**

video card: A **circuit board** that is mounted in one of the computer's **expansion slots** and provides processing and memory for video images. Turns a computer's digital signals into signals that the monitor can display on the screen.

video local bus: Video cards that use the **local bus**. Can also refer to the **bus** itself, or to the standards used to develop the card.

video RAM: RAM that is mounted on the **video card,** or sometimes on the **motherboard,** and kept separate from the system's **main memory.**

virus: A program generally intended to harm a computer by damaging or deleting data, software, or system files.

virus signature: A distinct piece of the code used to create a **virus,** which serves to identify the specific virus.

VL bus: See **video local bus.**

voice recognition: A technique that uses special software and a computer's **sound card** to allow the user to give the computer spoken commands, reducing dependence on keyboards, mice, and other manual devices.

voice synthesis: A technique which uses special software and a computer's **sound card** to pronounce out loud the contents of a text file, an error message, or other information.

volatile: Unable to retain information once power to a computer is cut off. Used to describe **RAM.**

WAN: See **wide area network.**

wide area network: Also **WAN.** A network covering a large geographic area, such as a country, state, or other large area. The Internet is the best example of a WAN.

windows terminal: A terminal that displays the output of a Windows-based program that is actually being run on a network server.

winsock: See **dialer software.**

Wintel: A nickname for the ubiquitous combination of the Windows **operating system** and **CPU** from Intel.

workstation: A high-end machine used for demanding tasks that computers with smaller capacity would be unable to do effectively.

WORM drive: Short for Write Once Read Many. Another term for **CDs** that can be written to once by the user, after which they are read-only.

writable: A disk onto which information can be written.

WYSIWYG: Acronym for What You See Is What You Get. Used to describe word processors or other editors that display a document just as it would appear on paper or over the Internet.

x86: The name of the family of **CPUs** that power most IBM-compatible computers. It refers to the original pattern of naming the chips, which started with the 8086 and progressed through the 80286, 80386, and so on.

XT bus: 8-bit version of the **ISA bus.** Appeared originally in the IBM PC/XT computer.

Bibliography

Batt, Chris. *Information Technology in Public Libraries.* 4th ed. London: Library Association, 1992.

Benson, Allen C. *The Complete Internet Companion for Librarians.* New York: Neal-Schuman, 1995.

Boschmann, Erwin. *The Electronic Classroom: A Handbook for Education in the Electronic Environment.* Medford, NJ: Learned Information, 1995.

Bucher, Katherine Toth. *Computers & Technology in School Library Media Centers.* Worthington, Oh.: Linworth, 1994.

———. *Information Technology for Schools.* 2nd ed. Worthington, Oh.: Linworth, 1998.

Davis, Frederic E. *Windows 95 Bible.* Berkeley, Calif.: Peachpit Press, 1996.

Desmarais, Norman. *CD-ROM Local Area Networks: A User's Guide.* Westport, CT: Meckler, 1991.

Derfler, Frank J., and Les Freed. *How Networks Work.* 2nd ed. Bestseller ed. Emeryville, CA: Ziff-Davis Press, 1996.

Engle, Mary, et al. *Internet Connections: A Librarian's Guide to Dial-up Access and Use.* Chicago: American Library Association, 1995.

Ensor, Pat. *The Cybrarian's Manual.* Chicago: American Library Association, 1997.

Fidelman, Miles R. *All-Out Internet Access: The Cambridge Public Library Model.* Chicago: American Library Association, 1997.

Fites, Phillip E., Peter Johnston, and Martin P. J. Kratz. *The Computer Virus Crisis.* New York: Van Nostrand Reinhold, 1992.

Harvey, Greg. *DOS for Dummies: Quick Reference.* Foster City, CA: IDG Books Worldwide, 1993.

Henle, Richard A., and Boris W. Kuvshinoff. *Desktop Computers in Perspective.* New York: Oxford University Press, 1992.

Jamsa, Kris. *Concise Guide to MS-DOS Batch Files.* Redmond, WA: Microsoft Press, 1994.

Judson, Jeremy. *The Macintosh Bible.* 6th edition. Berkeley, CA: Peachpit Press, 1996.

Keyes, Jessica. *The McGraw-Hill Multimedia Handbook.* New York: McGraw-Hill, 1994.

Ladner, Sharyn J. *The Internet and Special Libraries: Use, Training, and the Future.* Washington, D.C.: Special Libraries Association, 1993.

Landau, Ted. *Sad Macs, Bombs, and Other Disasters.* 2nd ed. Reading, MA: Addison-Wesley, 1995.

Liu, Cricket, et al. *Managing Internet Information Services.* Sebastopol, CA: O'Reilly and Associates, 1994.

Liu, Lewis-Guodo. *The Internet and Library and Information Services: Issues, Trends, and Annotated Bibliography, 1994–1995.* Westport, CT: Greenwood Press, 1996.

Louw, Eric, and Neil M. Duffy. *Managing Computer Viruses.* New York: Oxford University Press, 1992.

Lowe, Doug. *Networking for Dummies.* 2nd ed. Foster City, CA: IDG Books Worldwide, 1996.

MacDonald, Randall M. *The Internet and the School Library Media Specialist: Transforming Traditional Services.* Westport, Conn.: Greenwood Press, 1997.

McClure, Charles R., William E. Moen, Joe Ryan. *Libraries and the Internet/NREN: Perspectives, Issues, and Challenges.* Westport, CT: Meckler, 1994.

Minasi, Mark. *The Complete PC Upgrade and Maintenance Guide.* 8th ed. San Francisco, CA: Sybex, 1997.

Mueller, Scott. *Upgrading and Repairing PCs.* 8th ed. Indianapolis: Que, 1997.

Poole, Lon. *MacWorld Mac OS 7.x Bible.* Foster City, CA: IDG Books Worldwide, 1997.

Poole, Lon. *MacWorld Mac OS 8 Bible.* 6th ed. Foster City, CA: IDG Books Worldwide, 1997.

Ross, Calvin. *The Frugal Youth Cybrarian: Bargain Computing for Kids.* Chicago: American Library Association, 1997.

Schneider, Karen G. *The Internet Access Cookbook: A Librarian's Commonsense Guide to Low-Cost Connections.* New York: Neal-Schuman, 1996.

Simpson, Carol Mann. *Internet for Schools.* 2nd ed. Worthington, Oh.: Linworth, 1997.

Todino, Grace, John Strang, and Jerry Peek. *Learning the UNIX Operating System.* Sebastopol, CA: O'Reilly and Associates, 1993.

Valauskas, Edward J., and Monica Ertel. *The Internet for School Library Media Specialists: Today's Applications, Tomorrow's Prospects.* New York: Neal-Schuman, 1996.

Valauskas, Edward J., and Nancy R. John. *The Internet Initiative: Libraries Providing Internet Services and How They Plan, Pay, and Manage.* Chicago: American Library Association, 1995.

White, Ron. *How Computers Work.* Deluxe ed. Emeryville, CA: Ziff-Davis Press, 1997.

———. *How Software Works.* Deluxe ed. Emeryville, CA: Ziff-Davis Press, 1993.

Index

About the Author

Stephen Paling is the former technology services librarian at Emmanuel College in Boston and has worked in other areas of technology support and academic librarianship. He received his M.S. in Library and Information Science from Simmons College, where he is an adjunct faculty member. He is currently a doctoral candidate at Syracuse University.